NORTHROP'S T-38 TALON

A PICTORIAL HISTORY
Don Logan

Schiffer Military/Aviation History
Atglen, PA

ACKNOWLEDGEMENTS

I would like to thank the following individuals who have helped me in this project: Mr. C. John Armhein of Northrop/Grumman Corporation, Peter H. Becker, Tom Brewer, David F. Brown, Joe Bruch, George Cockle, John Cook, D.J. Fisher, Jerry Geer, Jim Geer, Norris Graser, Bob Greby, Steve Haskin, Roger Johansen, Renato E.F. Jones, Tom Kaminski, Craig Kaston, Ben Knowles, Robert L. Lawson, Ray Leader, Roy Lock, Kirk Minert, Hugh Muir, Brian C. Rogers, Mick Roth, Jim Rotramel, Douglas Slowiak, Keith Snyder, Terry Somerville, Keith Svendsen, Bruce Trombecky, Vance Vasquez, Ted Van Geffen, and Randy Walker.

THE AUTHOR

After graduating from California State University-Northridge with a BA degree in History, Don Logan joined the USAF in August of 1969. He flew as an F-4E Weapon Systems Officer (WSO), stationed at Korat RTAFB in Thailand, flying 133 combat missions over North Vietnam, South Vietnam, and Laos before being shot down over North Vietnam on July 5, 1972. He spent nine months as a POW in Hanoi, North Vietnam. As a result of missions flown in Southeast Asia, he received the Distinguished Flying Cross, the Air Medal with twelve oak leaf clusters, and the Purple Heart. After his return to the U.S., he was assigned to Nellis AFB where he flew as a rightseater in the F-111A. He left the Air Force at the end of February 1977.

In March of 1977 Don went to work for North American Aircraft Division of Rockwell International, in Los Angeles, as a Flight Manual writer on the B-1A program. He was later made editor of the Flight Manuals for B-1A #3 and B-1A #4. Following the cancellation of the B-1A production, he went to work for Northrop Aircraft as a fire control and ECM systems maintenance manual writer on the F-5 program.

In October of 1978 he started his employment at Boeing in Wichita, Kansas as a Flight Manual/Weapon Delivery manual writer on the B-52 OAS/CMI (Offensive Avionics System/Cruise Missile Integration) program. He is presently the editor for Boeing's B-52 Flight and Weapon Delivery manuals, B-1 OSO/DSO Flight Manuals and Weapon Delivery Manuals, and T-43A flight Manuals.

Don Logan is also the author of *Rockwell B-1B: SAC's Last Bomber* and *The 388th Tactical Fighter Wing: At Korat Royal Thai Air Force Base 1972* (both available from Schiffer Publishing Ltd.).

Book Design by Robert Biondi

Copyright © 1995 by Don Logan.
Library of Congress Catalog Number: 95-68024

All rights reserved. No part of this work may be reproduced or used in any forms or by any means – graphic, electronic or mechanical, including photocopying or information storage and retrieval systems – without written permission from the copyright holder.

Printed in China.
ISBN: 0-88740-800-1

We are interested in hearing from authors with book ideas on related topics.

Published by Schiffer Publishing Ltd.
77 Lower Valley Road
Atglen, PA 19310
Please write for a free catalog.
This book may be purchased from the publisher.
Please include $2.95 postage.
Try your bookstore first.

CONTENTS

INTRODUCTION 5

T-38 DEVELOPMENT 6

PILOT TRAINING T-38s 11
- USAF Pilot Training 11
- German Air Force Pilot Training 41
- EURO-NATO Joint Jet Pilot Training Program (ENJJPT) 43

FOREIGN T-38s 45

FLIGHT TEST SUPPORT/TEST PILOTS SCHOOL 47

AGGRESSORS 81

LEAD IN FIGHTER TRAINING (LIFT) 103

ACCELERATED CO-PILOT ENRICHMENT (ACE) 118

COMPANION TRAINER (CT) 119

THUNDERBIRDS 134

AVIATION INDUSTRY T-38s 139

AIRCRAFT DESCRIPTION/SYSTEMS 143
- Engines 143
- Fuel System 144
- Pressure Air System 144
- Air Frame Mounted Gearbox 144
- Electrical Systems 144
- Hydraulic Systems 144
- Flight Control System 144
- Wing Flaps 144
- Speed Brakes 144
- Landing Gear 144
- Ejection System 148
- Oxygen System 148
- Anti-G Suits 148
- Communication And Navigation Equipment 148
- AT-38B Special Equipment 148

Performance 151
Tail Codes 152
Colors & Markings 152

INTRODUCTION

The Northrop T-38A Talon, the only U.S. supersonic aircraft designed from the outset for pilot training, is completing its third decade with a record of effectiveness, safety and low cost unequalled in the history of supersonic flight. The T-38 is most widely known for its remarkable record in the Air Training Command. It continues to be used as the USAF Advanced Trainer for the new Air Education and Training Command, and has demonstrated its versatility in other roles for the Air Force, Navy, NASA, and civilian organizations. The Talon has been used in Aggressor roles; by the USAF Thunderbird Demonstration Squadron; support of Strategic Air Command missions as the U-2/SR-71 companion trainer; chase and test support for both Air Force and Navy test programs; As one of the aircraft in the Accelerated Co-Pilot Enrichment (ACE) program for Strategic Air Command; the Companion Trainer Program by Air Combat Command and Air Mobility Command; Engineering flight test operations at Wright-Patterson AFB, McClellan AFB and Kelly AFB; range support and photographic chase missions at Eglin AFB, and White Sands Missile Range; astronaut flight readiness training and other space program missions at Ellington AFB; NASA support at Langley AFB; and civilian avionics and flight test support.

T-38 DEVELOPMENT

In 1954 the Northrop Corporation saw the need for a jet trainer that would prepare pilots for the Mach 2-plus fighters, bombers and reconnaissance planes of the 1960s and 1970s. The new trainer must be lightweight, rugged, economical, and above all, safer than supersonic combat aircraft. It must have forgiving qualities that would allow inexperienced pilots to become safely adjusted to high sink rates, angles of attack and closure speeds. It must survive a training regimen that would require an average of 2 1/2 high-speed landings per flight hour. It must deliver supersonic performance, yet provide a much higher utilization rate than sophisticated combat aircraft. And, the new airplane must have low maintenance costs and an unprecedented service life.

Northrop had the answer in a lightweight fighter concept already under study. The T-38's development can be traced back to the N-156 lightweight fighter design program. The N-156 design incorporated a concept called "life cost." The term was originated by Tom Jones, Northrop's deputy chief engineer in 1954. "Life cost" means the total cost of a product over its entire life span. The goal was to have the lowest possible cost over the life of the aircraft. Northrop conducted a series of concentrated studies for a new type of manned weapon system. These investigations resulted in a totally new design philosophy. The future fighter had to incorporate certain key features: it had to be small; it had to be lightweight and low cost, with high thrust-to-weight engines; and it had to lend itself to simple maintenance.

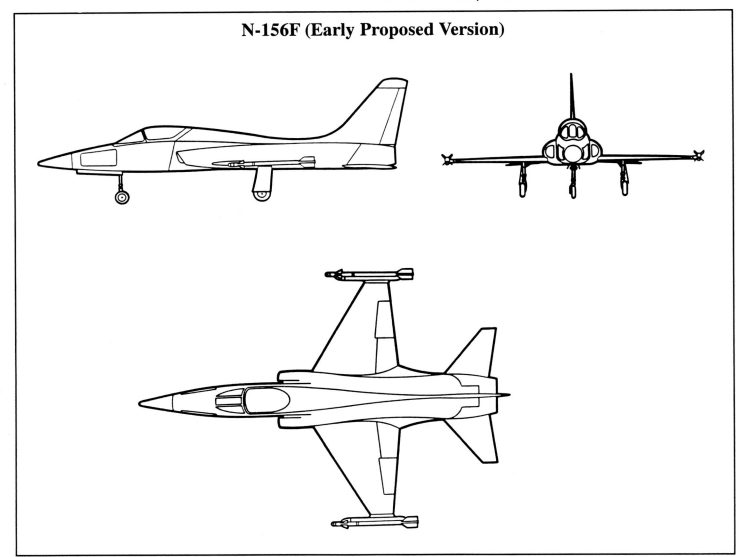

N-156F (Early Proposed Version)

N-156T (Early Proposed Version)

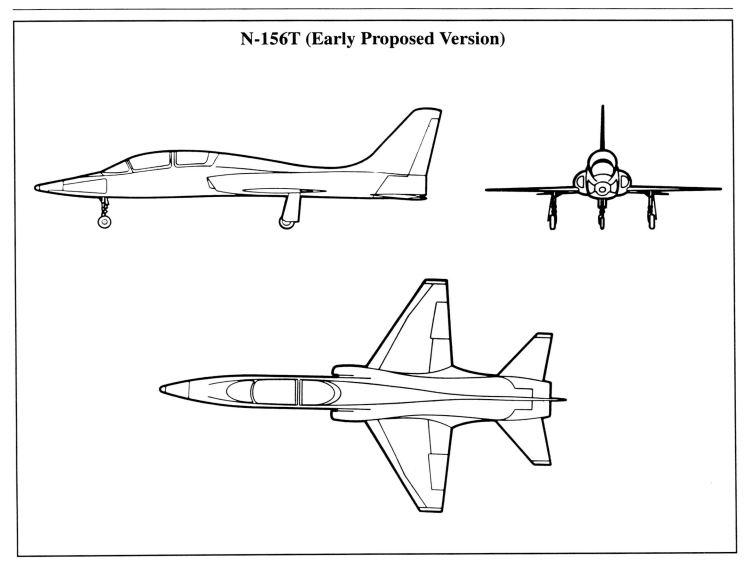

The timing was right because advanced lightweight high thrust-to-weight turbojet engine, the General Electric J-85, was under development for missile and target drone applications. Engineers recognized the potential of these engines for manned aircraft under study. The new engines were small enough so that two could be installed, increasing the aircraft safety factor. In fact, the two small-engine installation enhanced the weight savings compared to larger single-engine aircraft. The operational engine weighed only 584 pounds, yet delivered 3,830 pounds maximum thrust at sea level with afterburners. Two of these engines in an aircraft with a gross takeoff weight of approximately 10,000 pounds provided a formidable combination.

The first full airplane design to accept the advanced engines was the original N-156, a single-place fighter proposed for the Navy. It was developed by a design team headed by Welko Gasich. This short-coupled, T-tail, supersonic airplane of less than 10,000 pounds appeared to be ideal for operation from the small jeep carriers of the Navy. The proposal was short-lived, however, due to a Department of Defense decision to "mothball" the Navy's entire jeep carrier fleet.

In this same time period, a Northrop evaluation team made a tour of the allied nations of Europe and Asia to determine their future airpower needs. It was evident a firm requirement existed for an extremely versatile airplane, easy and inexpensive to operate and maintain, highly reliable, and with enough safety features that would result in a low attrition of both aircraft and pilot. A two seat version could also be used as a training aircraft.

In January 1955, Northrop made its first presentation to the U.S. government for a tactical supersonic airplane based on the "life cost" concept. The proposal included two versions: a single-place fighter designated N-156F, and a two-place trainer designated N-156T. This presentation coincided with the Air Force opening a competition for a supersonic "TZ trainer" to replace the T-33 subsonic jet then being used for Air Force advanced pilot training.

In mid-1955 the Northrop design team applied further refinements to the N-156F and N-156T designs, offering a number of briefings to U.S. and foreign military leaders. At the same time, the Air Force concluded that the N-156T met the "TZ trainer" specifications and issued a letter of intent to Northrop in June 1956 leading to development and production of the T-38.

The T-38 represented a major equipment change for the Undergraduate Pilot Training (UPT) program conducted by

The Number one YT-38 (58-1191), seen here on an early test flight had no rear cockpit, instead, the space was filled with instrumentation required for flight test. (Northrop Photo)

the Air Training Command (ATC). Replacing the subsonic T-33, it gave the student pilots a supersonic trainer with performance characteristics more closely corresponding to the Air Force's tactical fighters of the time. The T-38 with its light weight, high thrust engines was designed for transitional flight training in these tactical fighter categories: supersonic flight familiarization; takeoff and landing techniques; multi-jet engine operation; aerobatics; night flying; instrument instruction and cross-country navigation.

The June 15, 1956, Air Force letter of intent authorized development engineering for two prototype "YT" airplanes (58-1191 and 58-1192) and one static test airplane (58-1193). A series of low-speed wind tunnel tests was then conducted for aerodynamic refinement of the design, and a full-scale mockup was constructed under the designation TZ-156.

In October 1956, following a formal mockup board review, the Air Force approved the TZ-156 design. Tool design and manufacturing started in December, and the Air Force redesignated the airplane T-38. Production go-ahead was received that same month for an initial order for four pre-production T-38A airplanes (58-1194 – 58-1197). Quickly following was the first production order for 13 T-38As. The T-38s were built on the new NORAIL "moving assembly line" in 1958 at the Hawthorne facility. Production built up rapidly over the next two years.

Flight testing was divided into three categories. Categories I and II were USAF and Northrop design testing at Edwards AFB. Category III was an all-USAF evaluation of the aircraft under actual training conditions at Randolph AFB, Texas, headquarters of the Air Training Command.

The Number one YT-38 (58-1191) in its gloss white paint scheme was trucked from Hawthorne to Edwards Air Force Base where it began taxi tests in March 1959. The first flight took place on April 10, with Lew Nelson as test pilot. On the

Number one YT-38 first flew on April 10, 1959, after being trucked over 100 miles to Edwards AFB from the Northrop plant in Hawthorne, California. (Northrop Photo)

59-1594, the first production T-38, and 59-1197, the fourth production T-38A, taxi back from a test flight during June 1960. 59-1594 is in the normal two seat configuration, while 59-1197 is fitted with rear cockpit instrumentation similar to that in the prototype YT aircraft. (Via Craig Kaston)

third flight, four days later, with Nelson at the controls, the No. one YT-38 exceeded the speed of sound. No. two YT-38 (58-1192) made its first flight on June 12.

By the fall of 1959, the two prototypes had completed 100 flights, almost half of the entire flight test schedule. The T-38 had encountered none of the major problems usually experienced in flight testing supersonic aircraft; not a single aerodynamic change was required during the program. The T-38 was rapidly achieving a reputation for outstanding flight handling qualities with "forgiving" flight characteristics. Maintenance requirements were also proving to be very low.

Eight Air Force pilots flew each phase of the tests, using both YT-38s and five early production machines. Test instrumentation occupied the rear seat for a number of flights, the installed package including an oscillograph, photo-panel signal conditioning equipment and a camera. Production T-38s did have some features that distinguished them from the prototypes, including a curved metal plate over the exhaust nozzles, and an air intake on the fuselage top next to the vertical fin to cool the afterburners.

In October 1959, the Air Force contracted for 50 additional T-38 airplanes. The contract also included provisions for raising the tooling production capability rate from two to 10 aircraft per month. Ten months later, another order for 144 aircraft was received. In 1961, the production rate was increased from 10 to 12 airplanes per month.

The flight test program was completed in February 1961 following a total of 2,000 flights. Air Force officials announced the tests were the most successful ever conducted at Edwards AFB. The YT-38s and T-38As completed their total 8,588 hours of Category I, II, and III flight testing without a single accident – a record still unequalled in supersonic aircraft test programs. The T-38 proved its capabilities to ATC shortly after the first T-38 went to the Air Training Command at Randolph AFB,

59-1597, seen here in May 1961 is painted with the day-glow orange and "TF" aircraft identification number ("Buzz number") markings common on the early T-38s. (Via Craig Kaston)

A PICTORIAL HISTORY • 9

These two photos show the day-glow orange markings carried by the early T-38s. (Northrop)

Texas in the spring of 1961. In August, another 144 aircraft were ordered.

Other unprecedented events were to occur in the test program. At first, pilots were unable to put the T-38 into a spin. The spin test program had to wait until pilots and engineers could devise a special entry which consisted of rolling the aircraft onto its back and carrying out a precise series of control stick maneuvers. Ten years and nearly four million flight hours later, no one had yet put a T-38 into an unintentional spin. Even a vertical stall, a probable spin entry for most supersonic aircraft, would not cause a spin.

During the test program a contest to name the T-38A was conducted among USAF personnel. An instructor pilot's entry, Talon, was the winner.

On March 17, 1961 a sonic boom high over Randolph AFB announced the arrival of the first T-38A assigned to the Air Training Command. Col. Arthur F. Buck, the delivery pilot, then made a low-level pass before assembled Air Force and civic dignitaries and landed the aircraft for the presentation ceremony.

T-38 production continued at a steady rate as a result of incremental buys averaging 140 aircraft per year. In 1966 the Federal Republic of Germany purchased 46 Talons, and announced that it would conduct its entire student pilot program at Sheppard Air Force Base in Texas. The U.S. Navy joined T-38 users in 1969, and purchased five T-38s for the Navy Test Pilot School, Naval Air Test Center, Patuxent River, Maryland.

The Air Force accepted the 1000th T-38A (68-8095) on 7 January 1969, by which time ATC was using the aircraft at nine different bases. The tenth base to receive Talons was Columbus, Mississippi, with the 3650th Pilot Training Wing receiving its first of assigned 80 aircraft in the spring of 1969.

The final T-38 (70-1956) was delivered to the Air Force on January 31, 1972, at the Palmdale, California facility. It was assigned to the 3510 PTW at Randolph. Overall, a total of 1,189 airplanes and one static test airframe had been produced, 1,114 of which were for USAF Air Training Command.

Early in its career the Talon set numerous performance records. The renowned aviatrix Jacqueline Cochran piloted T-38As to eight world marks between August 24 and October 12, 1961, including a women's speed record of 844.2 miles per hour, a closed course distance record of 1,346.3 miles and a peak altitude record of 56,390.8 feet. In February 1962, Major Walter Daniel, USAF, set four world time-to-climb records in a Talon, reaching 3,000 meters in 35.62 seconds (including takeoff roll), 6,000 meters in 51.43 seconds, 9,000 meters in 64.76 seconds, and 12,000 meters in 95.74 seconds. Although these records have been surpassed, the Talon's climb rate is still one of the world's fastest, and student pilots and instructors find it useful in getting them to training altitude without delay.

SERIAL NUMBERS

58-1191		YT-38-05-NO Talon	(1)
58-1192		YT-38-05-NO Talon	(1)
58-1193		YT-38-05-NO Static Test Aircraft	(1)
58-1194	- 1197	T-38A-10-NO	(4)
59-1594	- 1601	T-38A-15-NO	(8)
59-1602	- 1606	T-38A-20-NO	(5)
60-0547	- 0553	T-38A-25-NO	(7)
60-0554	- 0561	T-38A-30-NO	(8)
60-0562	- 0596	T-38A-35-NO	(35)
60-0597	- 0605	T-38A Canceled Order	(9)
61-0804	- 0947	T-38A-40-NO	(144)
62-3609	- 3752	T-38A-45-NO	(144)
63-8111	- 8247	T-38A-50-NO	(137)
64-13166	- 13305	T-38A-55-NO	(140)
65-10316	- 10475	T-38A-60-NO	(160)
66-4320	- 4389	T-38A-65-NO	(70)
66-8349	- 8404	T-38A-65-NO	(56)
67-14825	- 14859	T-38A-70-NO	(35)
67-14915	- 14958	T-38A-70-NO	(40)
68-8095	- 8217	T-38A-75-NO	(123)
69-7073	- 7088	T-38A-80-NO	(16)
70-1549	- 1591	T-38A-85-NO	(43)
70-1949	- 1956	T-38A-85-NO	(8)
TOTAL			**(1,190)**

PILOT TRAINING T-38s

The Talon has successfully been used by the Air Training Command (ATC) as the advanced pilot training aircraft replacing the T-33 in Pilot Training Wings. The T-38 was ATC's first supersonic trainer. The first ATC T-38 (59-1595) was delivered to Randolph AFB on March 17, 1961. The Talon has been instrumental in graduating 75,000 student pilots since its introduction. It also achieved a very low accident rate, as compared to the overall Air Force rate. The T-38 required an average of only 10 man hours of maintenance per flying hour, or two-thirds of the originally projected rate, while maintaining a steady operational-ready rate of 80 percent.

USAF PILOT TRAINING BASES

The numerical redesignation of pilot training units occurred in 1972 when the wings were changed form Pilot Training Wings (PTW) to Flying Training Wings (FTW) and the numbers were changed from four digits to two digits.

T-38 PILOT TRAINING BASES/WINGS

BASE	ORIGINAL UNIT WING DESIGNATION	WING DESIGNATION AFTER 1972
Columbus AFB, MS	3650 PTW/3651 PTS	14 FTW, 50 FTS
Craig AFB, AL	3615 PTW	29 FTW, 52 FTS
Laredo AFB, TX	3640 PTW	38 FTW, 71 FTS
Laughlin AFB, TX	3646 PTW	47 FTW, 86 FTS
Moody AFB, GA	3550 PTW	38 FTW, 71 FTS (1973)
Sheppard AFB, TX	3630 FTW	80 FTW, 90 FTS
Randolph AFB, TX	3510 FTW	12 FTW, 560 FTS
Reese AFB, TX	3500 PTW	64 FTW, 52 FTS, 54 FTS
Vance AFB, OK	3575 PTW	71 FTW, 25 FTS
Webb AFB, TX	3560 PTW	78 FTW, 83 FTS
Williams AFB, OK	3525 PTW	82 FTW, 97 FTS

This photograph was taken from aircraft #4 of a Williams AFB, 82nd FTW, four-ship UPT training flight in February 1984. (Brian C. Rogers)

Until the mid-1970s, markings on T-38s varied little. All Pilot Training wings had the same basic markings with only a 12 inch ATC insignia and an AF Outstanding Unit Award ribbon above it. The serial number painted on the tail was in 12 inch high single stroke Gothic vertical lettering. (Don Logan)

The home base of a T-38 could sometimes be identified, while on a cross country flight, by the unit designation on the travel pod carried on the aircraft centerline – in this case the 12th FTW. The travel pod carried clothing and other necessities the pilots might need on an overnight stay away from home base. (Ray Leader)

Aircraft maintenance flights at some bases marked their aircraft with a colored stripe at the top of the tail. This colored stripe was later was changed to more specific unit markings. (George Cockle)

USAF Undergraduate Pilot Training (UPT)

By mid-1961 15 Talons had arrived at Randolph to take part in an intensive test and evaluation. The first ATC students to train in the T-38 came from Webb AFB – Class 62F. By year's end, Randolph had 44 new T-38s and Webb had 21.

By the end of 1963, as ATC continued to accept T-38s into its inventory, conversion was one year ahead of schedule with only Laughlin, Laredo, and Craig yet to begin conversion.

In 1965 the UPT syllabus was changed by adding the flight screening program which added the T-41. The previous program was a 55 week, 252 flying hour course. The new course, known as the 30/90/120 hour program decreased to 53 weeks with 240 flight hours (30 conventional and 210 jet). The T-41, an off the shelf Cessna 172, was used as the trainer for the new Flight Screening Program. Civilian instructors would provide 27 days and 30 flight hours of instruction. The goal of the flight screening program was to eliminate those candidates without the necessary skills to successfully complete pilot school. This screening was accomplished prior to flying the T-37, which was more expensive to operate.

The T-38 training syllabus included contact flying using visual ground references (flying in visual contact with the ground), two ship formation, instrument flying, navigation, and four ship formation. The contact phase of the syllabus included training in high performance flying skills (aerobatics, stalls, unusual attitudes, slow flight, and traffic pattern work).

In 1968 the DoD requested USAF help in training pilots for the U.S. Marine Corps. On June 21, 1968 Class 68-08 graduated from Laredo and Vance. In this class were the first group of USAF trained (including normal T-38 training) Marine Corps pilots. In conjunction with the cutback in pilot requirements following the end of the Vietnam conflict, Laredo Air Force Base was the first of several bases to give up the UPT mission as pilot production was dramatically scaled back in the mid-1970s. Laredo AFB closed on September 30, 1973, with the 38th FTW replacing the 3550th FTW at Moody AFB, Georgia. Following graduation of UPT class 76-04 on November 21, 1975, the 38 FTW was inactivated. On December 1, 1975, Moody AFB, Georgia transferred from ATC to Tactical Air Command (TAC).

During 1977, in an effort to reduce training costs, UPT was cut from its 210 flight hours to 170 hours, as flight simulators took the place of actual aircraft flight time for some training especially basic instruments. Two more bases, Craig and Webb were closed and the 29th FTW at Craig and the 78th FTW at Webb were inactivated. Craig graduated its last class (77-08) on August 12, and Webb graduated its last class on September 1, 1977. The T-38s available as a result of these base/wing closures were reassigned to the Lead In Fighter Program which dramatically expanded in 1977. Other T-38s, while still assigned to the FTWs were detached to SAC bases under the Accelerated Co-Pilot Enrichment (ACE) program which commenced in the mid-1970s. Lead In Fighter Training and Accelerated Co-Pilot Enrichment are discussed in greater detail later in this book.

In 1990 ATC reorganized its UPT wings (except the wing at Sheppard) following a trial program instituted at Williams AFB. The T-37 squadron and T-38 squadron were both split into two equal sized smaller squadrons. At Williams for example the 97th FTS had consisted of eight flights. Under the new program four flights stayed with the existing 97th FTS, and four flights moved to the new 99th FTS. A similar pattern occurred at Vance (26th FTS), Laughlin (87th FTS), Columbus (49th FTS) and Reese (52nd FTS).

In 1992, with the need for pilots steadily declining, the number of flying training squadrons in each wing was reduced from four (two T-37 and two T-38 squadrons) to two squadrons, with the 49th FTS, 86th FTS, and 26th FTS T-38 training squadrons being inactivated. The 52nd FTS at Reese remained active, but changed aircraft when they became the first flight training squadron to receive the T-1 Jayhawk. The 49th FTS at Columbus was reactivated and assigned the task of lead in fighter training, flying AT-38Bs.

As flight training wound down at Williams AFB, Arizona the 82nd Flying Training Wing inactivated on 31 March 1993, preparing for the base closure in the fall of 1993.

CURRENT AETC T-38 PILOT TRAINING BASES/WINGS

BASE	WING DESIGNATION	SQUADRON DESIGNATION	TAIL CODE
Columbus AFB, MS	14 FTW	49 FTS	CB
	50 FTS		
Laughlin AFB, TX	47 FTW	87 FTS	XL
Sheppard AFB, TX	80 FTW	88 FTS	EN
	90 FTS		
Randolph AFB, TX	12 FTW	560 FTS	RA
Reese AFB, TX	64 FTW	54 FTS	LB
Vance AFB, OK	71 FTW	25 FTS	OK

Specialized Undergraduate Pilot Training (SUPT)

In late 1987 the ATC began a number of policy changes which changed the training tracks and basing strategy for Specialized Undergraduate Pilot Training (SUPT). Where before there had been fighter-attack-reconnaissance and tanker-bomber-transport training tracks, now there were bomber-fighter and tanker-transport tracks. Before, fighter-attack-reconnaissance students, after completing UPT were assigned to Holloman AFB for Lead In Fighter training. In the new SUPT program all training would be provided at a single base. Reese was the first base programmed to offer SUPT, beginning in mid-1991. This program was called Specialized Undergraduate Pilot Training (SUPT) and continued after ATC's change to Air Education and Training Command (AETC), the name given to the reorganized Air Training Command. Under SUPT, each pilot training base would gain AT-38 fighter lead in trainers and T-1 Jayhawk trainers assigned in addition to the standard T-38s. Fighter lead in for USAF pilots would no longer be conducted at Holloman. All student pilots, after completing pilot screening at Hondo, Texas, flying the Slingsby T67M260 (USAF designation T-3A Firefly), would receive their primary pilot training in the Cessna T-37. During the T-37 training phase a determination is made, for each student, as to which type of aircraft they will be assigned upon graduation from UPT. The assignment options are transport type aircraft or fighter type aircraft (including fighters, attack, or bomber type aircraft). The students selected for transport assignments complete their pilot training in the Beech T-1 Jayhawk. Those going to fighter, attack, or bomber aircraft complete their training in the T-38 and AT-38 aircraft. The first SUPT class started training at Reese on July 20, 1992, and began T-1/T-38 training on January 24, 1993.

Mandatory Wing tail flashes were added to UPT units in the late 1970s. The 12th FTW's tail flash, as seen here on 62-14950 in September 1983, was a 12 inch high FS 15044 insignia blue stripe containing two rows of stars – four stars in the top row and five stars in the bottom row. (Ben Knowles)

The 12th FTW maintenance flights were numbered 1, 2, and 3. As seen here on 64-13209 in March 1985, the Flight number "1" was added to the top of the rudder in 12 inch high Gothic lettering. (Brian C. Rogers)

60-0554 seen here in July 1983 belongs to the #2 Flight of the 12th FTW as evidenced by the "2" on the rudder top. (Brian C. Rogers)

The 12th FTW's travel pods, like the one seen here on 68-8112 in January 1981, were painted gloss white with the Wing insignia applied to both sides. A commander or other senior officer's name was sometimes applied in two inch blue script lettering below the Wing insignia. (Brian C. Rogers)

The Wing Commander's aircraft normally used an aircraft whose serial number ended with the Wing's number. 68-8112, seen here in August 1983 has its serial number modified to serve as the 12th Flying Training Wing Commander's aircraft. (Brian C. Rogers)

66-8403, seen here in September 1984 belongs to the #3 Flight of the 12th FTW as evidenced by the "3" on the rudder top. (Brian C. Rogers)

A PICTORIAL HISTORY • 15

In the mid-1980s ATC experimented with paint schemes other than the standard gloss white. 70-1554, seen here in September 1984, was assigned to the 12th FTW and was painted in a blue and white camouflage scheme. In addition the Wing tail flash was removed and replaced with Tactical style tail codes, RA for Randolph AFB, home of the 12th FTW. (Steve Haskin Collection)

This photo of 70-1554 shows the camouflage scheme on the right side. (Craig Kaston Collection)

The paint scheme on 67-14946, seen here in May 1986, was changed slightly with light gray-blue patches added to the camouflage. (Bruce Trombecky)

A two tone scheme, adding a dark blue bottom and red lettering to the standard gloss white, was used at all the UPT bases. 68-8143, assigned to the 12th FTW, is seen here on October 9, 1987, taxiing out for a flight. (Brian C. Rogers)

As seen on 64-13209, the top of the wings and stabilizer remained gloss white. (Brian C. Rogers)

The 12th FTW Commander's aircraft, 68-8112, was also painted in the two-tone scheme. (Peter H. Becker)

A PICTORIAL HISTORY • 17

With the reorganization of the ATC as AETC on July 1, 1993 the UPT wing's markings changed. As seen here on 66-4333 on September 17, 1993, the unit tail stripes were retained, but tactical style tail codes and tactical style serial number presentation replaced the old tail numbers. (Brian C. Rogers)

Above and right: The markings for Commander's aircraft were also changed, with the unit number replacing the serial number. As seen here on 68-8124 on September 17, 1993, the serial number is not visible on the tail. (Brian C. Rogers)

Left and below: Additional levels of command received Commander's aircraft. Seen here on September 17, 1993, 70-1949 is marked as the Commander's aircraft for the 19th Air Force, the numbered air force to which the pilot training wings are assigned. (Brian C. Rogers)

Modifications to the serial number presentation are sometimes necessary. As an example, the aircraft seen in this photo, serial number 66-8357, should have the tail number 66-357, it has 8357 in large lettering instead of just 357 as the convention would dictate. This prevents confusion with aircraft 66-4357, also assigned to Randolph, which would also have 66-357 as its tail number. (Brain C. Rogers)

The 12th TFW has received its own AT-38Bs. The camouflaged jets, as here on 68-8109, carry the same RA tail code as the T-38s of the wing. (Don Mecili)

AT-38B 61-0851, mismarked as 62-0851, is now assigned to the 560th FTS at Randolph. (Don Mecili)

Another AT-38B 65-10437 is seen here at Randolph in February 1995. (Don Mecili)

Left and below: Markings on the travel pod identify 61-834 as a 14th FTW T-38 based at Columbus AFB. (Jerry Geer Collection)

During 1976 many U.S. military units applied special markings to individual aircraft in celebration of the Bi-Centennial. A 14th TFW marked 69-7076 is seen here. (Hugh Muir)

68-8200, seen here on August 8, 1987, assigned to the 14th FTW, painted in the two tone test paint scheme, carried CM tail. (Brian C. Rogers)

The cover attached to the inside of the rear canopy of 68-8200 is the instrument hood used to block visual cues from the student during instrument flight training. (Brian C. Rogers)

68-8204 is seen here in May 1992, with the tail stripe and intake covers of the 14th FTW. (Bob Greby)

Though the 14th Wing had previously used CM as their tail code, CB as seen here on 66-4339 was adopted for the present 14th FTW markings. (USAF)

Above and left: The Air Training Command (ATC) was redesignated the Air Education and Training Command (AETC) on July 1, 1993. The AETC insignia is seen here along with the CB tail code and tactical tail number on 68-8097 in May 1992. The tail stripe of the 14th FTW consists of a dark blue stripe edged in dark red with five stars in each edge stripe. Two chevrons, one white and one red are centered in the dark blue stripe. (Renato E.F. Jones)

Right and below: AT-38Bs of the 14th TFW, like 64-13203 seen here, carry the same markings as the 14th FTW T-38As. (Tom Kaminski)

AT-38Bs of the 14th FTW are assigned to 49th Flying Training Squadron. The 49 FTS Squadron Commander's aircraft is seen here. (Kieth Snyder)

Above and below: Aircraft of the 3651 Pilot Training Wing were painted in the standard ATC markings. Pilot training at Craig AFB ended on August 12, 1977, when the 3651 PTW graduated its last class. (Steve Haskin Collection)

65-10463 shot at Davis-Monthan was the 29th FTW Bi-Centennial aircraft. (Craig Kaston)

66-4359 an 47 FTW T-38 named "THE CITY OF DEL RIO" after Del Rio, Texas, the location of Laughlin AFB, is seen here on March 7, 1981. (Brian C. Rogers)

The 47th Flying Training Wing Commander's T-38, 68-8118, seen here at Carswell AFB Fort Worth Texas on October 7, 1985. The 47th FTW's tail stripe was dark blue containing four white T-38s flying in diamond formation. A small "XL" (excel) was centered above the stripe. (Brian C. Rogers)

The 47th FTW's maintenance flights were lettered U, V, W, X, and Y. A member of U flight, 65-10388, is seen here on May 6, 1985. (B.C. Rogers)

A member of V flight, 64-13195, is seen here on May 3, 1985. (Brian C. Rogers)

A member of W flight, 64-13224, is seen here on August 31, 1985. (Brian C. Rogers)

A member of X flight, 64-13254, is seen here on March 3, 1985. (Brian C. Rogers)

A member of Y flight, 63-8222, is seen here on March 15, 1985. (Brian C. Rogers)

A member of X flight, 66-4337, is seen here in two tone paint scheme with XL tail code on June 16, 1987 At least one other 47th FTW carried the LL tail code. (Brian C. Rogers)

The 47th FTW's tail stripe, reflecting things to come was changed to a plain dark blue stripe with XL in 12 inch white letters. (Douglas Slowiak/Vortex Photo Graphics)

Left: After the change over to AETC, the small XL became the Wing's tail code. The 87th FTS, which had been activated under the 47th FTW on April 2, 1990 replaced the 86th FTS. The tail stripe of was changed to the "Red Bulls", which reflected the 87th's heritage. The 87th FTS was previously the 87 Fighter Interceptor Squadron (FIS) from K.I. Sawyer AFB Michigan, flying the F-106 Delta Dart with the distinctive red bull's head painted on the tail. (Don Logan)

Below: 64-13242 is seen here on September 3, 1993, at Tinker AFB during a stop over on an out-and-back cross country flight from Laughlin. (Don Logan)

67-14951 taxiing at McChord AFB, Washington on June 17, 1994. (Renato E.F. Jones)

70-1584 carrying the Wing Commander's travel pod shows the tail stripe used on 64th FTW aircraft during the late 1970s. (Logan Collection)

65-10366 has a red stripe with a running fox, indicating the aircraft belongs to F (fox) flight of the 64th FTW. (Ben Knowles)

This 64th FTW T-38A, 62-3663, belongs to H (hotel) flight as indicated by "HOTEL" written in script in the blue tail stripe. (Brian C. Rogers)

Left and below: This 64th FTW T-38A, 61-849, belongs to I (India) flight as indicated by "INDIA" written in script in the tail stripe. (Ben Knowles)

As seen here on 70-1584 taken on November 16, 1980, the 64th FTW changed its tail stripe into a wedge during late 1980 and early 1981. (Ben Knowles)

62-3654, the 54 FTS Commander's aircraft, displays the 64th TFW tail stripe used during 1984 and 1985. This stripe contained a map of the state of Texas at the aft of the stripe. Each maintenance flight had its own color. (Brian C. Rogers)

In the late 1980s the 64th FTW changed its stripe once again. This change moved the map of Texas to the front of the stripe and replaced the remainder of the blue stripe with a red and a white stripe. The owning flight, in this case F (fox) flight, was identified by a 12 inch letter on the tail above the rudder. (Douglas Slowiak/Vortex Photo Graphics)

62-3652, the 52 FTS Commander's aircraft, belonging to G (golf) flight is seen here in November 1990. (Jim Geer)

This 64th FTW T-38, 63-3690, belongs to I (India) flight. (Brian C. Rogers)

Reese AFB and the 64th FTW adopted LB as its tail code as seen here on 66-4330. (Jerry Geer)

65-10379 is seen here with an LB tail code. LB stands for Lubbock, Texas, the location of Reese AFB. (Brian C. Rogers)

The 71st FTW aircraft used tail stripe color to indicate the owning flight. In addition, the stripe contained the letters OK standing for Oklahoma. Vance AFB the home of the 71st FTW is located in Enid Oklahoma. 62-3709 seen here in March 1994, has a red tail stripe. (Brian C. Rogers)

63-8113 seen here in January 1995, has a yellow tail stripe. (Brian C. Rogers)

63-8146 seen here has a black tail stripe. (Brian C. Rogers)

68-8171 is seen here, on September 12, 1992, marked as the 71st FTW Commander's aircraft with a red tail stripe. (Norris Graser)

The tail stripe was modified with the OK in black surrounded by white. 63-8127 seen here with a red tail stripe. (Jerry Geer)

The tail stripe on 65-10358 seen here is green. In the background is a 64th FTW with a red tail stripe. (Douglas Slowiak/Vortex Photo Graphics)

A PICTORIAL HISTORY • 35

Above: The 71st FTW adopted VN as its tail code. VN stands for Vance AFB. 70-1565, with the new tail code, is seen here on September 12, 1992. (Norris Graser)

Right: The tail markings on 63-8229 of the 71st TFW include the Tail Code VN. (Renato E.F. Jones)

Below: The present tail markings seen here on 70-1955 retained the red tail stripe with OK in black. (Norris Graser)

Left: The 78th FTW painted 70-1950 for the Bi-Centennial celebration. (Steve Haskin Collection)

Below: 62-3678 of the 78th FTW is seen here at Webb AFB on March 27, 1976. The 78th FTW graduated its last class on September 1, 1977, Webb AFB was closed soon after. (Steve Haskin)

The 78th FTW also used tail stripe colors to identify flights. Behind 62-3627 with a red tail stripe can be seen T-38s with blue, green, and red stripes. (Steve Haskin)

Three 82nd FTW T-38s (65-10320, 64-13198, and 64-13263) fly in formation over western Arizona. This photograph was taken from aircraft #4 of four-ship UPT training flight in February 1984. (Brian C. Rogers)

This 82nd T-38 (67-14928) photographed on June 28, 1980, carries a red and white checkered tail stripe. (Ben Knowles)

This 82nd T-38 (67-14848) photographed on February 22, 1984, carries a yellow tail stripe. (Brian C. Rogers)

This 82nd T-38 (68-8132) photographed on January 6, 1985, carries a blue tail stripe with "WILLIE" centered above the stripe. (B.C. Rogers)

The 82nd FTWs two tone T-38, 66-4389, photographed on September 26, 1987, carries a WL tail code. WL stands for Williams AFB. (Douglas Slowiak/Vortex Photo Graphics)

By 1985 the 82nd FTW, as seen here on 70-1563, had done away with the tail stripe and replaced it with an Indian head immediately above the tail number. (Douglas Slowiak/Vortex Photo Graphics)

A PICTORIAL HISTORY • 39

70-1582 seen here with "Willie One" marked at the top of the tail, This H flight aircraft was the 82nd FTW commander's aircraft in 1987. (Douglas Slowiak/Vortex Photo Graphics)

By 1991, the 82nds tail markings had changed again. The Indian head was now placed in a wide stripe at the top of the tail with "Willie" in script centered above the stripe. 66-4328 is seen here on July 27, 1991 with the new tail stripe. (Douglas Slowiak/Vortex Photo Graphics)

68-8217, seen here with the new tail stripe was the Wing Commander's aircraft during 1992. The 82nd FTW was inactivated on March 31, 1993. (Randy Walker)

GERMAN AIR FORCE PILOT TRAINING

Starting in October 1964, student pilots for the German Military began training at Sheppard AFB, Texas using T-37s and T-38s belonging to the USAF. Due to Europe's unpredictable weather disrupting training schedules, it was realized that there was a substantial advantage in basing the aircraft in the USA and having the pilots do their training there. In addition, areas of West Germany over which fast jet training could safely be conducted were very limited. As a result of these factors Luftwaffe jet training was transferred to America. In the beginning the German program used USAF aircraft. The Federal German Government approved the purchase of the T-38 and T-37 in July 1965 and the Defense Ministry ordered 47 T-37Bs for primary training and 46 T-38As for advanced jet training. The aircraft were financed in fiscal year 1966, the Talons included in a batch of 56 aircraft built as T-38A-65-NOs with serial numbers 66-8349 to 66-8404.

The first German aircraft were delivered to Sheppard AFB, under the operational control of the 3630th Pilot Training Wing organization of ATC. The first T-38, City of Wichita Falls, arrived in February 1967. They were not used exclusively by German students as each UPT class contained some USAF students. On April 21 the first group of USAF students began training with the German Air Force students. The instructor force at Sheppard was also a mix of German and USAF pilots. The GAF syllabus was 55 weeks long and consisted of 262 hours, 132 in the T-37 and 130 in the T-38, with no T-41 instruction. The Talons and T-37s at Sheppard were painted in USAF markings and integrated into the inventory of the 3630th PTW.

German Air Force UPT ended on August 7, 1982 when the last class graduated. Although designed primarily for German Air Force students, the course had also graduated students from the German Navy, Royal Netherlands Air Force, and the USAF. In all, the course produced 1,252 German, 49 Dutch, and 544 USAF pilots.

66-8380, was the 80th FTW Commander's aircraft in 1983. (Brian C. Rogers)

This T-38, 66-4352, with standard ATC markings was assigned to the 80th FTW for German Pilot Training. (Ben Knowles)

A PICTORIAL HISTORY • 41

Above and below: Sheppard AFB T-38s used for German pilot training, like 66-4346 seen here at Sheppard on April 18, 1974 had standard USAF ATC markings. (Don Logan)

Sheppard's two tone aircraft, 69-7073 seen here, had WF as the tail code for the 80th FTW. WF stands for Wichita Falls, Texas, the location of Sheppard AFB. (Brian C. Rogers)

EURO-NATO JOINT JET PILOT TRAINING PROGRAM (ENJJPT)

On May 17, 1978, ministers from the North Atlantic Treaty Organization (NATO) accepted an offer from the United States to host the Euro-NATO Joint Pilot Training program for a 10 year period starting in 1981. The 12 participating nations are; Belgium, Canada, Denmark, Germany, Greece, Italy, the Netherlands, Norway, Portugal, Turkey, the United Kingdom and the United States. On June 11, 1980 the Secretary of Defense announced that ENJJPT would be conducted at Sheppard. The German pilot training at Sheppard AFB since 1966 was changed to include training for the other NATO countries. The first class of ENJJPT students started training at Sheppard on October 1, 1981.

T-38s assigned to the 80th FTW for ENJJPT training have their tails marked with the NATO insignia. 66-8380, previously the German pilot training 80th FTW Commander's aircraft, remained the 80th FTW Commander's aircraft after the unit began training under the ENJJPT program. (Brian C. Rogers)

66-4374 became the 80th FTW Commander's aircraft after the transition to AETC. The 80th FTW adopted EN standing for Euro-NATO as its tail code. (Douglas Slowiak/Vortex Photo Graphic)

The 80th FTW also had a 90 FTS Commander's aircraft, seen here in May 1994. (Ben Knowles)

The 80th FTW aircraft are marked in the same manner as 63-8129 seen here. (Douglas Slowiak/Vortex Photo Graphic)

Two above: AT-38B 67-14842 is the 88 FTS Commander's aircraft. (Don Logan)

FOREIGN T-38s

In addition to the U.S. and Germany, other nations have operated T-38As. Portugal with 12 aircraft, Taiwan with 21 aircraft, and Turkey with 70 aircraft. They were all used for pilot training and transition into fighter aircraft. Sales of surplus USAF T-38As to other countries are presently under negotiations, with details to be released soon.

Portugal

61-0843
61-0853
61-0868
61-0890
61-0897
61-0915

Portuguese T-38 61-0868 seen here in September 1979 carries the USAF serial number. (Steve Haskin Collection)

Portugal replaced the USAF serial numbers with their own identification numbers. 2606 is the Portuguese identification number. (Craig Kaston Collection)

A PICTORIAL HISTORY • 45

64-13280 seen here is used by the 435th FS at Holloman for advanced pilot training of Taiwan's student pilots. (USAF)

Taiwan (Republic of China)

During 1977 and 1978 21 T-38s were shipped to Taiwan and used for advanced pilot training for those pilots scheduled to fly fighter aircraft. This program was canceled and presently the training of Taiwan's fighter pilots was moved to the U.S. with it being accomplished by the 435th Fighter Squadron of the 49th Fighter Wing flying AT-38Bs based at Holloman AFB.

T-38s Used In Taiwan During 1978

61-0810	61-0856	61-0905
61-0813	61-0858	61-0914
61-0819	61-0862	61-0916
61-0824	61-0865	61-0919
61-0830	61-0869	61-0926
61-0832	61-0887	61-0935
61-0838	61-0892	61-0937
61-0841	61-0893	61-0944
61-0850	61-0902	61-0946
61-0854		

On May 12, 1993 the 435th FS was activated to conduct fighter pilot training for the Taiwan Air Force. About 21 AT-38s were withdrawn from storage at AMARC for this mission.

AT-38s Presently Assigned To The 49th FW, 435th FS at Holloman AFB:

61-0807	61-0866	61-0940
61-0835	61-0875	62-3641
61-0845	61-0876	62-3678
61-0847	61-0880	62-3752
61-0852	61-0891	63-8164
61-0860	61-0899	64-13280
61-0863	61-0911	68-8140

During March and April of 1995 the following 40 T-38As were shipped by boat to Taiwan for use in the pilot training program of Taiwan's Air Force:

61-0819	62-3654	63-8113	63-8155
61-0849	62-3657	63-8116	63-8156
61-0918	62-3665	63-8118	63-8177
61-0925	62-3674	63-8126	63-8184
61-0936	62-3689	63-8127	63-8230
61-0945	62-3701	63-8138	63-8241
62-3623	62-3717	63-8139	64-13194
62-3625	62-3720	63-8143	64-13216
62-3626	62-3730	63-8144	64-13231
62-3629	62-3750	63-8150	64-13253

Turkey

The Turkish Air Force operates 70 T-38s in a pilot training role. 30 aircraft were delivered in the first batch with an additional 43 aircraft being delivered in September of 1993.

T-38As Delivered In The First Batch

62-3611	62-3728	63-8201
62-3617	62-3737	63-8203
62-3621	62-3649	63-8205
62-3624	63-8145	63-8206
62-3627	63-8159	63-8208
62-3649	63-8161	63-8210
62-3688	63-8173	63-8216
62-3696	63-8183	63-8220
62-3708	63-8191	63-8231
62-3711	63-8195	63-8233

During 1993, 43 additional T-38As were shipped by air to Turkey. The first 40 were shipped in C-5s, eight per C-5, in five different flights. The last three were shipped in Turkish Air Force C-130s, one per C-130 flight.

C-5 Shipment 1 February 22, 1993	C-5 Shipment 3 June 8, 1993	C-5 Shipment 5 October 25, 1993
62-3667	62-3718	62-3644
62-3675	62-3744	62-3734
63-8176	63-8157	62-3740
63-8237	63-8209	63-8137
64-13178	63-8226	63-8180
64-13200	63-8227	63-8238
64-13207	63-8240	64-13221
64-13223	64-13237	64-13236

C-5 Shipment 2 April 22, 1993	C-5 Shipment 4 August 24, 1993	C-130 Shipment 1 October 21, 1993
		62-3748
62-3619	62-3630	
62-3656	62-3693	C-130 Shipment 2 November 6, 1993
62-3661	62-3719	
62-3713	62-3729	62-3743
62-3751	62-3739	
63-8115	63-8121	C-130 Shipment 3 November 20, 1993
64-13180	63-8151	
64-13300	63-8190	62-3721

FLIGHT TEST SUPPORT/TEST PILOTS SCHOOL

The T-38 was used in many support roles requiring regular use of aircraft by the armed services test centers and NASA. Routine duties included flying as the chase plane on prototype test flights. In 1969 an order for five was placed by the U.S. Navy for the Test Pilot School at NATC Patuxent River, Maryland. The Air Force Test Pilot School at Edwards AFB also uses T-38s in its program.

AIR FORCE MATERIEL COMMAND (AFMC) T-38s

AFMC, formerly Air Force Systems Command (AFSC) operated T-38s at its test facilities; Edwards AFB, California, Eglin AFB, Florida, Kirtland AFB New Mexico, White Sands Missile Range (Holloman AFB), and the various bases within the Nevada Test range complex. In addition, T-38s were oper-

58-1197, the fourth production T-38A seen here over the Mojave Desert north of Edwards AFB, was used in flight test work at Edwards for many years after the T-38 test program had been completed. (USAF)

59-1597 is seen here at China Lake Naval Weapons Center with the USAF Test Pilot School insignia on the tail. (Ben Knowles)

ated at a number of AFMC's (formerly Air Logistics Command), Air Logistics Centers (ALC) including; Ogden ALC (Hill AFB, Utah), Sacramento ALC (McClellan AFB, California), and San Antonio ALC (Kelly AFB, Texas).

Edwards AFB Flight Test Support

Following the completion of T-38 Flight Test at Edwards AFB, T-38s remained assigned to the Air Force Flight Test Center (later the 6510 Test Wing, and now designated 412th Test Wing) performing training and support duties. The following T-38s were assigned to the 412th TW as of 31 March 1994

61-0825	67-14943	70-1558
63-8135	67-14956	70-1559
65-10325	68-8153	70-1574
65-10375	68-8154	70-1575
65-10402	68-8158	70-1579
67-14856	68-8205	

This insignia of the Air Force Systems Command (AFSC) has been carried by many T-38s over the years. (Don Logan)

The Air Force Flight Test Center (AFFTC) aircraft carried a blue stripe with white Xs as seen here in June 1981 on 61-0825. (Brian C. Rogers)

67-14954 is seen here in 1976 with the Bi-Centennial insignia displayed between the stripe and the tail number. This insignia was carried by many U.S. Military aircraft during 1976. (Hugh Muir)

67-14943, assigned to the 6512 Test Squadron of the 6510 Test Wing at Edwards AFB, seen on January 3, 1985 flying over the foot hills of the Sierra Nevada Mountains, north of Edwards. (Keith Svendsen)

65-10402 taxis back to its parking spot at Edwards AFB. The aircraft was assigned to the 6510 Test Wing when this photo was taken on November 9, 1986. (Craig Kaston)

Four T-38 tails displaying the AFFTC insignia are seen here on the Edwards flight line. (Craig Kaston)

A PICTORIAL HISTORY • 49

Right and below: Edwards AFB tail code ED, along with the Air Force Materiel Command (AFMC) was added to the Edwards T-38s after AFMC was established on July 1, 1992 as part of the reorganization of the USAF. (Don Logan)

Sitting on the Edwards flight line on August 14, 1994, 70-1575 awaits a flight crew. (Don Logan)

Eglin AFB Flight Test Support

The Air Force Development Test Center (46th Test Wing) based at Eglin AFB, Florida carries an ET (Eglin Test) tail code. This unit was formerly the Air Force Armament Test Center (ADTC) 3246th Test Wing. Eglin AFB units have operated the following T-38s in support of weapons test programs.

Eglin AFB T-38s

61-0834	70-1566
61-0844	70-1567
61-0861	70-1570
61-0874	70-1571
70-1557	

63-8187, seen here on June 2, 1973, assigned to the Air Force Development Test Center (AFDTC) aircraft based at Eglin Air Force Base Florida, carries a tail stripe of red diamonds. (Ben Knowles)

AFDTC T-38 65-10352, seen here on November 4, 1978 still carries the Bi-Centennial insignia on its tail, in addition to the AFSC insignia behind the cockpit. 65-10352 later flew with NASA as NASA 910. (Ray Leader)

65-10326, seen here on August 18, 1977, carries the Armament Division insignia and tail stripe of red diamonds. (Ben Knowles)

The AD tail code as seen here on 70-1570 on April 8, 1983 was carried by aircraft of the 3246 Test Wing of AFDTC at Eglin until replaced by ET in 1989. (Craig Kaston Collection)

ET (Eglin Test) tail code is seen here on 70-1570 in June 1991. (Craig Kaston)

46th Test Group (TG)/6585 Test Group

The 46th Test Group (previously designated the 6585 Test Group), 586th Test Support Squadron at Holloman AFB flies two AT-38Bs in support of testing on the White Sands Missile Range (WSMR). These AT-38s carry an HT (Holloman Test) tail code. The 46th Test Group is assigned to the 46th Test Wing based at Eglin AFB Florida.

T-38s operated in support of WSMR

62-3660   63-8215   70-1558

Above: 70-1558 assigned to Holloman AFB is seen here at Williams AFB on October 10, 1974, in the standard T-38 gloss white paint. (Don Logan Collection)

Left: The two AT-38s of the 6585th TG at Holloman AFB, like 63-8215 seen here March 11, 1993, carried HT tail codes. (Vance Vasquez)

63-8215 seen on a stop over at Point Mugu. (Vance Vasquez)

The 6585th TG was redesignated the 46th Test Group, and the tail of 63-8215 seen at Holloman on April 1, 1994, changed to reflect the new designation. (Ben Knowles)

As seen on April 1, 1994, 62-3660 carried the markings of the 586th Test Support Squadron which was assigned to the 46th TG. (Ben Knowles)

Kirtland AFB Flight Test Support

The Air Force Special Weapons Center (AFSWC) at Kirtland AFB, New Mexico operated T-38s in support of nuclear weapons test programs.

Kirtland AFB T-38s

61-0908	70-1566
61-0930	70-1567

70-1567, at Kirtland AFB on June 14, 1974 displays the Air Force Special Weapons Center (AFSWC) insignia and the tail stripe of alternating red and yellow triangles. (Ben Knowles)

61-0908 and 61-0930 seen here on the ramp at Nellis AFB during a stop over in April 1975. (Don Logan)

4477th Test and Evaluation Squadron (TES)

The 4477th TES Red Eagles' operated T-38s during the 1984-1988 time period in support of USAF testing of Soviet aircraft acquired by the U.S. This testing was accomplished in the Nevada Flight Range complex north of Las Vegas.

Serial Number	Camouflage	Nose Number
60-0553	Special Lizard	53
60-0572	3 Grays	72
	2 Blues	72
61-0870	3 Grays	70
	2 Blues	70
61-0851	3 Blues	51
68-8106	3 Blues	06

60-0553, while assigned to the 4477th TES Red Eagles, was painted in the three shades of brown "Special Lizard" camouflage. This photo was taken on November 15, 1987. (Douglas Slowiak/Vortex Photo Graphic)

68-8106, a former Thunderbird aircraft and the newest of the 4477th's T-38s, seen here on June 10, 1988, was painted in the Red Eagles' three blue camouflage scheme. (Brian C. Rogers)

60-0572 seen here on December 8, 1985 in the Red Eagles two blue paint scheme. (Brian C. Rogers)

60-0572 as seen here on June 10, 1988, was re-painted in the Red Eagles three gray camouflage scheme. (Brian C. Rogers)

This photo taken on November 15, 1987, shows the three gray pattern on the left side of 60-0572. (Douglas Slowiak/Vortex Photo Graphic)

This photograph taken on March 3, 1985, shows 61-0870 painted in the Red Eagles two blue paint scheme. (Brian C. Rogers)

61-0870, was also re-painted in the Red Eagles three grays. (Brian C. Rogers)

4450th Tactical Group (TG)

The T-38s of the 4450th Tactical Group flew in support of the F-117 program and operated out of Tonopah Test Range Air Field in Nevada. The aircraft served as chase and companion trainer aircraft. Since no two seat F-117 were available, the T-38 being flown by a F-117 instructor pilot would fly chase during the first training sorties of new F-117 pilots. The aircraft came from the 4477th TES and the 479th TTW. They were painted a distinctive three gray camouflage scheme, with TR tail codes. The group was later replaced by the 37th Tactical Fighter Wing, later 37th Fighter Wing, keeping the TR tail code.

3 Gray Camouflage

Light Gray   16440
Gray   16473
Dark Gray   16251

4450 TG/37 TFW T-38s

60-0553	61-0848	65-10367
60-0572	61-0851	

60-0572 along with other 4477 aircraft was transferred to the 4450 Test Group carrying TR tail codes, and was re-painted in the three gray camouflage scheme carried by the 4450th. (Ben Knowles)

60-0572 is seen here at Nellis during August 1990 after the 4450th had been redesignated as the 37th Tactical Fighter Wing. (Ben Knowles)

65-10367, seen here in June 1992 at Nellis AFB, was also assigned to the 37th TFW. (Ben Knowles)

Differing from the 37th T-38s, 61-0851 did not have "37 TFW" painted on the tail, as seen here in July 1990. (Ben Knowles)

60-0553, previously the "Special Lizard" 4477th T-38 is seen here in August 1992 painted in the 37th's three gray camouflage. (Ben Knowles)

Ogden Air Logistics Center (OO-ALC) Flight Test Support, Hill AFB

The 6514th Test Squadron, (later the 514th TESTS and then the 514 FLTS) now the 15th Flight Test Squadron (FLTS) based at Ogden Air Logistics Center has operated T-38s in support of their activities as the Air Logistics center for F-4s and F-16s. The following T-38s have been assigned to Hill AFB:

OO-ALC/Hill AFB T-38s

60-0569   61-0861   61-0913

60-0569 while assigned to Hill AFB carried the high visibility red and white paint scheme. In addition to the red on the nose and tail, the outer wing panels and the stabilators were painted red. (Ben Knowles)

60-0569 was operated along with F-4s at Hill AFB by the 6514th Test Squadron. This photo was taken on June 21, 1973. (Ben Knowles).

Sacramento Air Logistics Center (SM-ALC) Flight Test Support, McClellan AFB

The 337th Flight Test Squadron (FLTS), which replaced the 2874 TESTS in October 1992, operated T-38s at SM-ALC. Initially T-38s were assigned directly to Sacramento Air Logistics Center and operated in support of activities of the center including the F-111 flight test. 60-0551 was assigned to Sacramento Air Logistics Center. 60-0551, previously used by Northrop as a factory demonstrator, was used by Jacqueline Cochran to set eight women's world flight records, four for speed, two for distance, and two for altitude. The aircraft is presently on display in the air museum located at McClellan AFB.

60-0551 was operated by the Sacramento Air Logistics Center at McClellan AFB. This view of 551 was photographed at Nellis AFB in February 1975. (Don Logan)

60-0551, seen here at McClellan AFB on September 8, 1976, had the Bi-Centennial insignia added to the fuselage during 1976. (Craig Kaston)

By 1982 the paint scheme on 60-0551 had been changed to reflect its new task as part of F-111 engineering flight test. (Craig Kaston)

A PICTORIAL HISTORY • 63

In 1987 60-0551 was assigned to the 2874 Test Squadron. The F-111 aircraft previously on the tail has been changed to a T-38. (Craig Kaston)

San Antonio Air Logistics Center (SA-ALC) Flight Test Support, Kelly AFB

The 313th FLTS based at San Antonio Air Logistics Center has operated T-38s in support of their activities as the Air Logistics center for F-5s and T-38s. 65-10376 was assigned to San Antonio Air Logistics Center.

65-10376, seen here at Kelly AFB on October 12, 1976, was assigned to San Antonio Air Logistics Center. (Steve Haskin Collection)

Air Force Plant 42, Palmdale, California

Air Force Plant 42 operated T-38s in support of the SR-71 program at Palmdale, California.

Palmdale T-38s

63-8204   65-10363

Right: 63-8204, ex-NASA 904, seen here on October 14, 1974, still wearing the NASA stripe, was assigned to Air Force Plant 42 at Palmdale, California to serve as chase aircraft for SR-71 test flights. (Ray Leader)

Two below: 65-10363 seen here on February 2, 1980, replaced 63-8204 as the Palmdale SR-71 chase aircraft. (Mick Roth, Ben Knowles)

9th Strategic Reconnaissance Wing (SRW)

The 9th SRW at Beale operated a number of T-38s in support of their SR-71 and U-2 operations. Additional discussion of their present use of T-38s by the 9th Reconnaissance Wing is contained in the Companion Trainers chapter.

9 SRW T-38s

60-0578	64-13217	64-13277
60-0581	64-13225	64-13278
62-3653	64-13235	64-13281
62-3699	64-13240	64-13293
64-13182	64-13247	64-13297
64-13190	64-13262	64-13301
64-13194	64-13270	64-13302
64-13201	64-13271	64-13304
64-13212	64-13272	

Above and below: This "FLAGSHIP" of the 9th Strategic Reconnaissance Wing, 64-13271, photographed in June 1979, was one of a number of T-38s used in support of SR-71 operations at Beale AFB. (Ben Knowles)

The 9th SRW T-38's carried the SAC insignia on the left side of the tail as seen here on 64-13271 on May 31, 1979. The 9th SRW insignia was on the right side of the vertical tail. (Don Logan Collection)

By 1981 a yellow tail stripe containing four black iron crosses, taken from the Wing insignia as seen here on 64-13270, had been added to the tail. (Brian C. Rogers)

By March 1988 the 9 SRW had added a yellow and black Sabre stripe to the fuselage of their T-38s, as seen here on 64-13281 with the name "WHITE NINJA" below the windshield. (Brian C. Rogers)

Seen here on June 6, 1991, 64-13271 now has 9 SRW in large letters on the tail. (Douglas Slowiak/Vortex Photo Graphic)

U.S. NAVY Flight Test Support

Some of the early production T-38s have been converted to QT-38s and used by the Navy at China Lake Naval Weapons Center as target drones.

The Navy installed an AIM-9 capable missile launch rail on the centerline of 60-0582 (designated NT-38, then T-38B). The aircraft was also used in the first tests of a helmet mounted sight.

U.S. NAVY TEST PILOT SCHOOL
NAS PATUXENT RIVER, MARYLAND

USAF SERIAL NUMBER	NAVY BUREAU NUMBER	REMARKS	NOSE No.
59-1604			
60-0582			11
		Only AIM-9 capable T-38 (designated NT-38A-later T-38B)	
61-0855			
61-0889			11
63-8200			11
65-10327		Ex NASA 907	14
68-8194	158197	Crashed Nov 16, 1972	
68-8209	158198		10
68-8212	158199	Crashed Oct 24, 1974	
68-8214	158200		14/200
68-8216	158201		

68-8214 was given the USN Bureau Number 158200 assigned to USN Test Pilot's School, seen here on June 19, 1973. (Craig Kaston Collection)

A PICTORIAL HISTORY • 69

68-8216 (Bureau Number 158201) seen here on September 20, 1975, with a navy blue and gold TPS tail cap. (Ray Leader)

158201 seen here in June 1981 with a white tail cap. (Don Logan Collection)

65-10327, which also was NASA 907, is seen here during September 1978 in USN TPS markings. (D.J. Fisher)

Left: 70-1575's tail in October 1974 showing the USN Test Pilot School insignia below a non standard marking. (Don Logan Collection)

Below: 70-1575, later assigned to Edwards AFB, is seen on May 20, 1974 with a USAF markings with a USN Test Pilot School tail. (Don Logan Collection)

63-8200 was photographed at USN TPS in December 1976 was still wearing the NASA stripe left from when it had been NASA 903. (Bruce Trombecky)

59-1604 seen here on October 7, 1985, at USN TPS, still wearing its TOP GUN camouflage. (Ray Leader)

USN TPS experimented with the a two tone paint scheme similar to ATC's two tone. 68-8209 is seen here on October 5, 1987 wearing the Navy's version of dark blue and white. (David F. Brown)

60-0582 displays TPS's latest paint scheme, still carried on TPS's T-38s. (Don Logan Collection)

U.S. NAVY NAVAL WEAPONS CENTER (NWC) CHINA LAKE

SERIAL NUMBER	MODEL	DISPOSITION	NOSE NUMBER REMARKS
58-1194	QT-38A	Probably Shot Down	
58-1195	QT-38A	Probably Shot Down	
59-1594	QT-38A	Storage at NWC	#13
59-1595	QT-38A	Landing accident NAS Miramar 1 Oct 74	#17
59-1596	QT-38A	Probably Shot Down	#12
59-1597	QT-38A	Shot Down Dec, 1977	Orange
59-1598	QT-38A	Shot Down Dec, 1975	#16
59-1600	QT-38A	Storage at NWC	#14 White/Orange
59-1603	QT-38A	Probably Shot Down	#15, #381, #546
59-1604	T-38A		#11, #547
61-0851	T-38A	Transferred to 4477th TES - 1983	#551

59-1596 was China Lake's Bi-Centennial aircraft. (USN, Don Logan)

59-1594 seen at Nellis AFB in November 1975. The rear cockpit seat has been removed and replaced with flight test instrumentation painted orange. (Don Logan)

QT-38A 58-1195, the second production T-38A, The two blade antennas on the back of the aircraft identify it as a QT-38. (Don Logan Collection)

QT-38A 59-1600 is seen here in red-orange paint. (Don Logan Collection)

This T-38 is seen at China Lake in October 1975 still in its TOP GUN camouflage. (Don Logan)

T-38A 61-0851 on the flight line on May 10, 1980, at China Lake, in markings of the Naval Weapons Center Commander Captain William Neff. (Don Logan Collection)

59-1603 wearing its unique camouflage with the NWC Eagle on the tail, was photographed at China Lake during April 1988. (Don Logan)

A PICTORIAL HISTORY • 75

NASA

In May 1964 the National Aeronautics & Space Administration accepted the first of 30 T-38s which were to be used primarily to maintain the flight proficiency of astronauts. Many of the astronauts had already flown the T-38 as part of their military pilot training and two, Neil Armstrong and Elliot See, had been involved in the development program for the Talon and F-5. Armstrong was one of the NASA pilots flying early F-5 tests, and See had previously been a T-38 project test pilot with Northrop.

Astronaut Ted Freeman was killed in a 1964 T-38 (63-8188) crash when his aircraft crashed after hitting a large goose. Elliot See was killed, with fellow astronaut Charles Bassett, in a T-38 (63-8181) crash early in 1966. The astronauts departed from Ellington AFB near the Johnson Space Center, Houston, flying with another NASA T-38, to visit the McDonnell plant in St. Louis. See's aircraft ran into snow and fog at St. Louis, stalled on a go-around from the aborted landing approach, and hit one of the McDonnell Douglas factory buildings. Both crew members were killed in the crash. In 1967 Clifton Williams died after he lost aileron control in his T-38 (66-8354), and hit the ground at Mach 1 plus.

In NASA service, the T-38s all wore three digit numbers on their blue trimmed, white paint scheme. They supplemented T-33s and other types for training and general flight work, and are also used by astronauts who often have to commute long distances across the USA.

Recently several surplus AT-38s have been transferred to NASA.

NASA is in the process of converting some of its T-38As into what is unofficially being called T-38Ns. The "N" model has special cockpit modifications including a color weather radar, electronic flight instrument system, electronic communications control, and a navigation control and display unit.

NASA T-38s

NASA NUMBER	USAF SERIAL NUMBER
—	63-8188
	Crash October 31, 1964, Fatal
67	66-8354
511	65-10329
511	65-10330
717	65-10357
821	65-10353
901	63-8181
	Crash February 28, 1966, Fatal
901	66-8381
902	63-8193
903	63-8200
904	63-8204
905	65-10326
906	65-10326
907	65-10327
907	61-0912
908	65-10328
909	65-10351
910	65-10352
912	65-10354
913	65-10355
914	65-10356
915	65-10357
915	60-0586
916	66-8382
917	66-8383
918	66-8384
919	66-8385
920	66-8386
921	66-8387
922	66-8354
	Crash October 5, 1967, Fatal
923 (T-38N)	66-8355
924	67-14825
955 (T-38N)	69-7082
956	69-7084
957	69-7086
958	69-7088
959	70-1550
960 (T-38N)	70-1552
961 (T-38N)	70-1555
962	70-1556
963	59-1603
964 (AT-38)	68-8113
965 (AT-38)	68-8116
966	65-10357

In this July 19, 1972, photo NASA 924 (67-14825) is flying formation with 67-14949 in ATC markings. (USAF)

NASA's first markings included "NASA" in block letters, the NASA number, and the NASA blue circle insignia. (NASA)

NASA's markings changed, placing a yellow stripe and "NASA" above the number, and moving the blue insignia below the number. A blue "racing" stripe was added running the full length of the fuselage. The NASA registration number was painted in white in this stripe between the trailing edge of the wing and the horizontal stabilizer. (Brian C. Rogers)

NASA 909 (65-10351) was used as a chase aircraft for testing the Space Shuttle Enterprise. Noteworthy are the enlarged speedbrakes, painted with red diagonal stripes, which allowed the T-38 to fly the steep descent angle of the Space Shuttle without accelerating away from the shuttle. (Don Logan Collection)

NASA 955 (69-7082) received the new NASA tail markings consisting of only the modernized NASA script and identification number. (NASA)

On the Edwards AFB runway, NASA 919 (66-8385) accelerates for takeoff. (NASA)

A tail stripe of red diamonds left over from its assignment to AFDTC at Eglin AFB is seen on NASA 910 (65-10352) in this shot taken June 2, 1981. (Brian C. Rogers)

Left: In this photo NASA 904 (63-8204) is flying chase for NASA 947, the Grumman Gulfstream III used by astronauts to simulate Shuttle approaches for proficiency flight training. (NASA)

Below: NASA has once again changed their tail markings. 963 is (59-1603) seen here in July 1994 with the full tail NASA insignia and 63 at the top of the tail. (Jerry Geer)

NASA has begun modification of some of their T-38s into what is unofficially being called a T-38N. The "N" can be identified by its black radome. (NASA)

A PICTORIAL HISTORY • 79

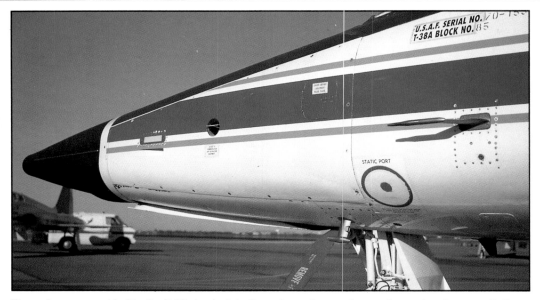

The radome was added to the T-38s to contain the antenna for a color weather radar. As a result, the pitot boom, previously on the nose was relocated to the side of the fuselage below the serial number block, as seen here on NASA 960 (70-1552).

These photos of NASA 960, the front cockpit on the left and the rear cockpit on the right, show the "glass cockpit" installation which includes an EFIS (Electronic Flight Instrumentation System) in the center with an electronic communication control and display unit below it, and control and display unit for the navigation functions on the left above the pilots knee.

AGGRESSORS

As a result of a low kill ratio during the U.S. involvement in Vietnam, it became obvious that more intensive training in air-to-air combat with dissimilar aircraft. The aircraft would replicate the performance of Soviet Bloc fighters. U.S. Air Force and Navy fighters flying against the North Vietnamese Air Force MiGs barely averaged a 2:1 kill ratio. This was nowhere near as high as that of Korea with an air superiority figure of approximately 10:1 in the U.S. favor. It was therefore decided that both the Navy and the Air Force would develop training programs which would give U.S. pilots the upper hand in air-to-air combat.

A unique concept was developed. For training purposes, U.S. pilots would fly "enemy" aircraft. Aggressor Squadrons would be set up with the objective of simulating enemy tactics using aircraft resembling Soviet bloc types which U.S. air forces had met, and presumably would continue to meet, over a battlefield. The pilots flying these aircraft would plan, fly and fight like Soviet bloc air forces.

During investigations into this plan, it became obvious that potentially the most dangerous aircraft to be faced by U.S. pilots in third world air force inventories at that time (mid-1970s) would be the MiG-21. The MiG-21 was therefore the primary aircraft to simulate in the aggressor scenario. Its features, particularly small size in comparison to contemporary Air Force and Navy fighters, should ideally be matched as closely as possible. Only the T-38/F-5 series came anywhere close in size and performance of the MiG-21.

Good as the T-38 was in establishing the parameters for TOP GUN and the Aggressors, it was lacking in the required performance and both services needed something a little "hotter." It was logical to use the F-5E, but few were available in the early 1970s. Northrop's production of F-5s at that time was devoted to MAP orders, and there were no plans to increase production. After the fall of South Vietnam, 77 F-5Es already programmed for the South Vietnamese air force under MAP became available. These F-5s found their way into U.S. service and replaced the T-38s in aggressor roles. Most of the Air Force aggressor T-38s were reassigned to Holloman AFB where they began new careers as Lead-In Fighters.

VF-43 adversary T-38 61-0913, is seen here flying formation with an F-14A (Bureau Number 159457) during a training mission flown out of MCAS Yuma on January 15, 1977. (Robert L. Lawson)

U.S. NAVY AGGRESSORS

In 1969, the U.S. Navy acquired six T-38As (former Edwards AFB jets) for evaluation as a suitable aggressor aircraft. The Talons, received by the Navy, were initially based at NAS Miramar with VF-121, the Pacific Fleet Readiness Air Group. Experienced Vietnam combat veterans developed a weapons and tactics course for the fleet Phantom II crews which started in March 1969 as the U.S. Navy Post Graduate Course in Fighter Weapons Tactics and Doctrine. The aggressor aircraft received special paint schemes which set them apart from other aircraft in the U.S. inventory.

One week was spent on air-to-ground weapons employment and three weeks were centered on air-to-air combat. The course was attempting to compress far too much into four weeks. The air-to-air phase demanded more time, and since Navy Phantom II crews were rarely involved in air to ground combat, the air-to-ground phase was dropped. The air combat phase was subsequently increased to five weeks. Soon after the course started it was given the name TOP GUN. The T-38s remained in service at Miramar into the period when the TOP GUN course became autonomous in 1972. The TOP GUN T-38s continued in use for two more years until the first F-5s were introduced by the Navy and began to replace the T-38s at TOP GUN.

The adversary program got underway with four T-38As and gradually worked up to operational proficiency by 1975. Home based at NAS Oceana, Virginia, the VF-43 Detachment initiated the Fighter Wing One ACM Readiness Program (FFARP). VF-43 would deploy against operational units and play the "enemy" during combat training.

NAVY AGGRESSOR T-38s

USAF SERIAL NUMBER/ NAVY BUREAU NUMBER	NOSE NUMBER	USING AGENCY
59-1594	13	Top Gun - NWC
59-1595	14	Top Gun
59-1596	12	Top Gun - NWC
59-1598	16	Top Gun - NWC
59-1600	—	Top Gun - NWC
59-1603	15/546	Top Gun - NWC
59-1604	547	Top Gun - NWC
59-1605	11	Top Gun
59-1606		
60-0582		VF-43 - VF-126
61-0851	551	Top Gun - 851 NWC
61-0855	552	Top Gun - Navy TPS
61-0882	4	VF-43
61-0913	3	VF-43
61-0915		VF-43
61-0918		VF-43
61-0929		VF-43

Below: T-38A 59-1594 "TOP GUN 13" in TOP GUNs three blue camouflage photographed at Naval Air Station (NAS) Miramar in February 1974. (Roy Lock)

T-38A 59-1596 "TOP GUN 12" in the TOP GUN desert camouflage photographed at NAS Miramar in June 1974. (Ted Van Geffen)

T-38A 59-1603 "TOP GUN 15" in the TOP GUN three gray camouflage photographed on November 18, 1974 at NAS Miramar. The "US" on the main gear door hints that bottom of the aircraft may be all white with full sized markings. (Ben Knowles)

T-38A 59-1598 "TOP GUN 16" seen with a replacement rear canopy still in white paint. (Don Logan Collection)

T-38A 59-1595 "TOP GUN 17" seem here on June 21, 1974, shortly after its arrival at NAS Miramar, is being prepared for repainting. (Don Logan Collection)

"TOP GUN 546" (previously "15") seen here on January 17, 1976 with TOP GUNs new markings which included the USN Fighter Weapons School insignia of a MiG-21 centered in a gunsight. (Don Logan)

The Navy Fighter Weapons School experimented with a number of camouflage schemes on the T-38. 61-0851 is seen here at NAS Miramar on January 17, 1976, painted in two blue wrap-around scheme. (Don Logan)

59-1604, seen here on January 17, 1976, was one of the first aircraft to be painted in a splinter camouflage. (Don Logan)

A PICTORIAL HISTORY • 85

"TOP GUN 552" 61-0855 seen here on July 13, 1976, is painted in light gray. (Don Logan)

By October 1977 gold blocks for crew member names had been added to 552. (Don Logan)

Seen here on February 19, 1978, all the markings on 552 have been painted over with a medium gray paint. (Don Logan)

VF-43 Challengers "4" 60-0882 seen here on an aggressor mission being flown out of MCAS Yuma. (Robert L. Lawson)

Above: 60-0582, seen here in May 1982 while assigned to VF-43 was the only T-38 capable of launching an AIM-9. The missile could be launched from a centerline missile rail using an optical sight. This aircraft was also used in early testing of a helmet mounted sight system. (Tom Kaminski)

Left: A VF-43 four-ship, an A-4F, two T-38s and a TA-4J, based at NAS Oceana, Virginia, in flight over Kitty Hawk, North Carolina. (USN)

Right: An F-5E adversary of VF-43 leads a T-38, A-4F, and TA-4J flying out of NAS Oceana, Virginia. (USN)

Below: 61-0882 on the ramp at NAS Oceana on September 26, 1976. (D.J. Fisher)

VF-43 "3" 61-0913 on a adversary mission over southern Arizona. (Robert L. Lawson)

Left: T-38 741 of VFA-127 was stationed at NAS Lemoore California. (Steve Haskin Collection)

Below: 60-0582 VF-126 "50" seen on July 25, 1987 in VF-126 markings still carrying its AIM-9 centerline launcher. (Brian C. Rogers)

59-1504 carried an unusual two gray camouflage when photographed on May 5, 1983. (Brian C. Rogers)

USAF AGGRESSORS

The USAF's own study into improved air combat training began in 1970. The evaluation to find a suitable aircraft to fulfill the aggressor mission was undertaken by the 64th Fighter Weapons Squadron at Nellis AFB, Nevada. As with the Navy program, the T-38 was found to be the only type in USAF inventory which could adequately duplicate the MiG-21's flight envelope. The 64th FWS initially received 20 Talons on loan from Air Training Command and went operational on 1 June 1973.

As with the Navy T-38s, the Air Force Talons were given new paint schemes; each one sported a distinctive camouflage scheme and later had Russian-type ID numbers added. The Air Force chose special names for their camouflage patterns including Ghost, Grape, Lizard, Snake, and Gloss Gray. These schemes were designed to match those used by Soviet bloc air forces. They took some time to perfect, and the mix did not always come out as planned. Pilots were only too aware of this when they were scheduled to fly a T-38 in a color scheme which contrasted with the terrain over which the mission was being conducted.

RED FLAG broadened the aggressor concept to train every TAC fighter squadron and beyond; TAC would send their units to Nellis to take part in RED FLAG flying on the Nellis ranges, practicing combat as a unit, with the pilots passing on their new found skills to other members of the squadron. RED FLAG 75-1 was held at Nellis in November 1975. The T-38 was an integral part of every RED FLAG. The 64th FWS pilots constituted Red Force, which Blue Force (the visitors) have to beat, not only in air combat, but by successfully attacking ground targets – dummy radar sites, missile batteries, airfield, tanks and vehicle convoys-dispersed throughout the ranges.

The T-38s began being replaced with F-5Es in late 1976, and many of the aggressor T-38s went to Holloman AFB to be used in lead in fighter training. Most were later modified to the AT-38 configuration.

USAF AGGRESSOR PAINT SCHEMES

GLOSS GRAY	Aircraft Gray 16473
GHOST	Blue 35237, Blue 35622, Gray 36307, Gray 36251
GRAPE	Blue 35414, Blue 35109, Blue 35164, Blue 35622
SNAKE	Brown 30118, Yellow 33531, Green 34256
LIZARD	Brown 30118, Yellow 33531

64th FWS T-38s

Serial Number	Camouflage	Nose Number	Disposition
61-0835	White	–	AT-38
61-0863	White	–	AT-38
61-0866	White	–	AT-38
62-3614	Gloss Gray	64	AT-38
62-3632	Grape	32	AT-38
62-3660	Lizard	60	AT-38
62-3666	Gloss Gray	66	
62-3673	Light Blue/ Gray Test	–	
	Grape	73	AT-38
63-8214	Gloss Gray	14	AT-38
63-8247	Lizard	47	AT-38
64-13168	Desert Test	–	
	Snake	68	
64-13169	Lizard	69	AT-38
64-13172	Lizard	72	AT-38
64-13188	Gloss Gray	88	AT-38
64-13192	Snake	92	
64-13193	Gloss Gray	93	AT-38
64-13276	Ghost	76	AT-38
64-13280	Gray Test	–	
	Ghost	80	AT-38
65-10370	Grape	70	AT-38
65-10382	Ghost	82	AT-38
65-10399	Snake	99	AT-38
65-10450	Snake	50	AT-38
65-10466	Grape	66	AT-38
67-14842	Ghost	42	AT-38

64-13168 here on the ramp at Nellis AFB showing the temporary desert camouflage pattern on the right side. (Steve Haskin Collection)

In early 1974, these four T-38s were painted with water based paints to test camouflage schemes for the USAF aggressor program. 64-13168 was painted in the desert scheme of tan, beige, and gray. 64-13280 was painted in two shades of gray. 62-3673 was painted in solid light blue-gray. 65-10466 was painted in light blue-gray and two shades of gray. (Steve Haskin Collection)

Right: 64-13280 at Nellis showing the two gray pattern on the left side of the aircraft. (Steve Haskin Collection)

Below: 64-13280, in March 1974, was repainted overall light gray using water based paint. (Ben Knowles)

The water based paint didn't last long. As seen in this photo taken in late May, the paint had already started to flake from the tail. (Steve Haskin Collection)

Left: 65-10466 seen in March 1974 with its test camouflage. (Steve Haskin Collection)

Below: 62-3673 was temporarily painted light blue-gray. This photo was taken at Nellis in March 1974. (Ben Knowles)

The aggressor T-38s were delivered to 57th Fighter Weapons Wing in the standard gloss white paint scheme. 61-0835 is seen here on December 30, 1974, shortly after arrival at Nellis. The 57th FWW yellow and black checkerboard tail stripe and TAC insignia were added after their arrival. (Don Logan)

64-13188 was painted gloss gray (16473) here at Nellis on December 30, 1974. (Don Logan)

Nose numbers consisting of the last two numbers of the serial number were added during 1975. 64-13188, in keeping with this convention, numbered 88 seen here on an aggressor mission in November 1976. (Don Logan)

62-3614 was an exception to the nose number rule. As the Squadron Commander's aircraft for the 64th Fighter Weapons Squadron, it was numbered 64 instead of 14. (Don Logan)

Left: The aggressor camouflage schemes were given code names. The Ghost scheme was a four tone camouflage of blue-gray (FS 35237), duck egg blue (FS 35622) light sea gray (FS 36307) and aggressor Gray (FS 36251). 67-14842 seen here at Nellis on February 10, 1975 shortly after a fresh Ghost camouflage had been applied. (Don Logan)

Below: 64-13276 during August 1975 is seen in Ghost scheme with its 76 nose number. (Don Logan)

65-10382 was also painted in the Ghost scheme. (Don Logan)

Grape camouflage scheme was a four blue scheme of FS 35414, FS 35109, FS 35164 and FS 35622. 62-3632 is seen here in the Grape scheme on April 4, 1975. (Don Logan)

65-10466, seen on August 12, 1975, in the Grape scheme, with 66 as its nose number. (Don Logan)

65-10370, shown here in Grape scheme, on January 27, 1976 flying an aggressor mission with its wing in primer, had recently completed a wing modification to increase its service life. (Don Logan)

Lizard camouflage was a two tone scheme of field drab brown (FS 30118) and tan (FS 33531). 62-3660 is seen on April 24, 1975, with a fresh Lizard scheme. (Don Logan)

64-13169, with nose number 69, seen on July 22, 1975, had its Lizard scheme modified by touching up the scheme with dark brown. (Logan)

64-13172 is seen here on July 22, 1975 with its nose number 72. (Don Logan)

Snake scheme was a modification of Lizard camouflage. It was a three tone scheme adding green (FS 34256) to a field of drab brown (FS 30118) and tan (FS 33531). 65-10450 is seen here on April 24, 1975, with a fresh Lizard scheme. (Don Logan)

64-13168, with nose number 68, was photographed at Nellis during August 1975. (Don Logan)

Another Snake was 64-13192, seen here at Nellis on July 22, 1975. (Don Logan)

65-10399, seen here on August 12, 1975, in the Snake scheme. (Don Logan)

Though the T-38 was replaced as an aggressor by the F-5E, a few T-38s remained at Nellis. 61-0870 seen here on July 4, 1980, was painted in an experimental three gray scheme. (Kirk Minert)

This view shows the opposite side of 61-0870. "70" was later assigned to the 4477 TES Red Eagles. (Kirk Minert)

26th TFTAS T-38s

A second USAF aggressor unit, the 26th Tactical Fighter Training Aggressor Squadron, 3rd Tactical Fighter Wing, was established at Clark AB, The Philippines. The 26th TFTAS first eight T-38As arrived at Clark AB on January 23, 1976. By June all eight had been painted in aggressor camouflage schemes. The 26th operated a total of 13 T-38As before being replaced by F-5Es in 1978. Of the 13 aircraft, two had crashed, seven were returned to Davis-Monthan AFB and put in storage, and four remained at Clark AB until the end of 1980 when they too were sent to D-M. Due to extensive corrosion which occurred during their stay at Clark, the aircraft were sold for scrap.

26th TFTAS T-38s

Serial Number	Camouflage	Nose Number	Disposition
65-10330	Gloss Gray	30	Returned to DM 1978
65-10362	Lizard	62	Returned to DM 1978
65-10365	Grape	65	Returned to DM 1978
65-10389	Gloss Gray	89	
	Ghost	89	Remained with 26th until 12/80
65-10391	Unknown	?	Crashed October 20, 1976
65-10400	Ghost	00	Remained with 26th until 12/80
65-10406	Snake	06	Remained with 26th until 12/80
65-10409	Ghost	09	Remained with 26th until 12/80
65-10411	Grape	11	Returned to DM 1978
65-10426	Grape	26	Returned to DM 1978
65-10441	Lizard	41	Returned to DM 1978
65-10443	Unknown	?	Crashed March 30, 1977
65-10448	White	None	Returned to DM 1978

65-10400, in the Ghost scheme with 00 nose number, was photographed at Clark AB in February 1978. (Jerry Geer Collection)

65-10389 painted in all gray at Clark AB in February 1978. (Jerry Geer Collection)

By November 1979 when this photograph was taken, 65-10389 had been repainted in Ghost camouflage. (Steve Haskin Collection)

65-10809, seen here at Clark AB in November 1979, was one of three T-38s painted in the Ghost scheme. (Steve Haskin Collection)

65-10406 is seen here at Clark AB in the Lizard scheme. (Steve Haskin Collection)

The pattern of the Snake scheme used by the 26th TFTAS differed from that of the 64th FWS Snake scheme, as seen here on 65-10441. (Steve Haskin Collection)

LEAD IN FIGHTER TRAINING (LIFT)

The fighter lead in program was initiated in early 1975 and was set up at Holloman AFB with the 465 Tactical Fighter Training Squadron (TFTS) of the 49 TFW operating standard T-38s for the program. After completing UPT, new pilots scheduled for assignment to tactical aircraft were sent to Holloman AFB for fighter lead in training. An earlier USAF program used AT-33s at Nellis AFB, Cannon AFB, and Myrtle Beach AFB for training new fighter pilots. It was determined early in the program the LIFT aircraft should have a limited air to ground and air to air weapons capability. As a result, although the changes made to the T-38 throughout its lifetime were not to be extensive, Northrop did develop additional capability for the USAF's foremost trainer by making provision for it to carry weapons. With very limited U.S. procurement of the F-5B and only small numbers of them available to MAP training program, it was logical to give the Talon some "teeth" to enhance the realism for training fighter pilots.

The first T-38 to be fitted with a weapons store was 60-0576, part of the third production batch. This aircraft was redesignated LIF T-38 or T-38B, Northrop's Lead In Fighter. An ejector rack could be attached to the pylon, which was situated under the second cockpit, and was stressed to take the SUU-20A rocket/practice bomb carrier, a practice bomb rack or a SUU-11A/A minigun pod. Other T-38s in small numbers from various production batches, including most of the 64th FWS Aggressor T-38As, were subsequently modified by Air Force TCTO 1T-38A-889 into as T-38B Lead In Fighter trainers. A total of 134 T-38As were modified. The T-38B aircraft were subsequently redesignated as AT-38Bs.

LEAD IN FIGHTER TRAINING AT HOLLOMAN

Lead in Fighter Training began at Holloman in 1975 with the 465 TFTS, 49 TFW flying standard T-38s.

The 479th Tactical Training Wing (TTW) was activated on January 1, 1977 with three Tactical Fighter Training Squadrons (TFTS) assigned; 434th TFTS, 435th TFTS, and 436th TFTS. The 416th TFTS was added on March 15, 1979. The 433rd TFTS replaced the 416th TFTS on September 1, 1983. The squadrons, except the 433rd, were inactivated during the first half of 1991.

On November 15, 1991 Holloman reorganized under the "One Base – One Wing" concept. The single remaining AT-38B flying squadron, the 433rd FS, was reassigned to the newly formed 49th Operation Group (OG) which was assigned to the 49th FW on November 15, 1991. The 49th FW was the F-15 wing stationed at Holloman. The 479th FG was inactivated on the same date.

On July 8, 1992 the Holloman squadrons were restructured in conjunction with the arrival of the F-117s and the departure of the F-15s. The 433rd FS was inactivated and its AT-38s along with its personnel became the 8th FS. In September of 1992 the 8th FS was split into two squadrons with the 7th FS commencing AT-38 operations.

On July 1, 1993, Holloman again reorganized its squadrons. The 8th FS and the 9th FS replaced the 415th and 416th as the F-117 squadrons, with the 7th FS continuing as the lone remaining USAF AT-38 LIFT squadron. On December 1, 1993 Holloman ceased USAF LIFT training and the 7th FS

Holloman AFB can be seen in the distance behind these two 49th TFW T-38s. (Jim Rotramel)

became an F-117 squadron. The 7th FS continued to operate a small number of T-38s as companion trainers. Lead in fighter training has since been integrated into the UPT syllabus at the pilot training bases.

FIGHTER LEAD-IN UNITS
Holloman AFB

WING

49th TFW	LIFT initiated in early 1974 LIFT transferred to the 479th TTW on January 1, 1977
479th TTW	Activated January 1, 1977 Inactivated July 26, 1991
479th FG	Activated July 26, 1991 Inactivated November 15, 1991

479th TACTICAL FIGHTER TRAINING SQUADRONS

416th TFTS	HM	Silver	March 15, 1979 - September 1, 1983
433rd TFTS	HM	Green	September 1, 1983 - July 8, 1992 Redesignated 433 FS on November 1, 1991, and then reassigned to the 49th Operations Group (OG) on November 15, 1991.
434th TFTS	HM	Red	January 1, 1977 - May 3, 1991
435th TFTS	HM	Blue	January 1, 1977 - February 15, 1991
436th TFTS	HM	Yellow	January 1, 1977 - August 2, 1991

AT-38B Camouflage   Blue 35450,
Blue 35164,
Blue 35109

49th FIGHTER SQUADRONS

7th FS	HO	Gloss Black with light gray markings, used as F-117 companion fighter.
8th FS	HO	AT-38 Blue camouflage scheme, used for lead in fighter training when it replaced the 433rd TFTS on July 8, 1992.
433rd TFTS	HO	AT-38 Blue camouflage scheme, used for lead in fighter training after the 479th TTW was inactivated.
435th FS	HO	AT-38 Blue camouflage scheme, used for Taiwanese Air Force fighter pilot training.

During 1993, 1994, and 1995 lead in fighter training was transferred to AETC and accomplished at the UPT bases. The units presently performing lead in fighter training are:

BASE	WING	SQUADRON
Columbus AFB, MS	14th FTW	49th FTS
Laughlin AFB, TX	47th FTW	87th FTS
Sheppard AFB, TX	80th FTW	88th FTS
Randolph AFB, TX	12th FTW	560th FTS
Reese AFB, TX	64th FTW	54th FTS
Vance AFB, OK	71st FTW	25th FTS

AT-38B's

60-0550	61-0864	63-8117	65-10337
60-0553	61-0866	63-8149	65-10341
60-0561	61-0875	63-8153	65-10346
60-0569	61-0876	63-8162	65-10350
60-0572	61-0878	63-8164	65-10367
60-0573	61-0880	63-8166	65-10370
60-0576	61-0886	63-8172	65-10371
60-0589	61-0888	63-8175	65-10381
60-0591	61-0891	63-8187	65-10382
60-0594	61-0895	63-8207	65-10399
60-0595	61-0898	63-8211	65-10403
61-0804	61-0899	63-8214	65-10425
61-0806	61-0904	63-8215	65-10432
61-0807	61-0907	63-8247	65-10437
61-0809	61-0911	64-13169	65-10439
61-0812	61-0917	64-13172	65-10450
61-0814	61-0923	64-13188	65-10452
61-0817	61-0938	64-13193	65-10456
61-0818	61-0940	64-13203	65-10457
61-0820	61-0947	64-13211	65-10466
61-0828	62-3614	64-13215	65-10472
61-0831	62-3627	64-13232	67-14842
61-0835	62-3632	64-13245	68-8106
61-0836	62-3641	64-13261	68-8109
61-0842	62-3660	64-13264	68-8113
61-0845	62-3673	64-13267	68-8116
61-0847	62-3678	64-13269	68-8123
61-0848	62-3703	64-13276	68-8133
61-0851	62-3715	64-13280	68-8138
61-0852	62-3738	64-13288	68-8140
61-0857	62-3746	64-13292	68-8142
61-0860	62-3752	64-13298	68-8168
61-0863	63-8112	65-10321	

Left: While assigned to the 49th TFW, the T-38 carried the wing insignia on both sides of the aircraft spine aft of the rear cockpit. (Don Logan)

Below: T-38s of the 465 TFTS carried a brown tail stripe with six yellow check marks, as seen here on 60-0550. (Jim Rotramel)

For a short time, the 465th TFTS T-38 carried the HO tail code of the 49th TFW. (D.J. Fisher)

A PICTORIAL HISTORY • 105

Right: The prototype AT-38 lead in fighter, 60-0576 in a vertical climb over Edwards AFB. (USAF via Steve Haskin)

Below: 60-0576 is flying north of Edwards accompanied by 60-0550 from the 49th TFW and 67-14856 from the 6510 Test Wing at Edwards. (USAF via Steve Haskin)

The AT-38 was modified to carry practice weapons suspension units on a newly installed centerline pylon. 60-0576 is seen here carrying the SUU-11A/A during testing at Edwards AFB. (USAF via Steve Haskin)

60-0576 is seen here at the Edwards AFB open House in November 1975, as well as the weapons equipment it can carry: a SUU-20 Bomb Dispenser, an AF/B37K-1 Bomb Container, and a SUU-11A/A Mini-gun pod. (Don Logan)

60-0576 seen during March 1976, after its arrival at Holloman. (Douglas Slowiak/Vortex Photo Graphic)

This AT-38, 64-13232 is assigned to the 416th TFTS of the 479th TTW. The tail band for the 416th is silver. (Brian C. Rogers)

AT-38 63-8187 assigned to the 433rd TFTS, 479th TTW taxis for takeoff at Holloman. This 433rd TFTS AT-38 carries the unit's green tail band. (Don Logan)

This 434th TFTS aircraft, 61-0831, a T-38A was photographed on October 26, 1979 at Davis-Monthan AFB. It carries the red tail band of the 434th TFTS. (Ben Knowles)

61-0806, photographed on January 12, 1980, can be identified as belonging to the 435th TFTS by the blue tail band. (Ben Knowles)

This AT-38, 60-0561, is carrying a SUU-20 loaded with six BDU-33 practice bombs. The yellow tail band identifies it as a 436th TFTS aircraft. (Ben Knowles)

Though belonging to the 434th TFTS of 479th TTW, this AT-38, photographed in March 1978, carries the HO tail code of the 49th TFW. (Douglas Slowiak/Vortex Photo Graphic)

A PICTORIAL HISTORY • 109

Right: This 434 TFTS AT-38, 64-13193, an ex 64th FWS aggressor carrying a tactical travel pod on its centerline, is seen here at Nellis on June 21, 1980. (Ben Knowles)

Below: T-38A 64-13280, still in the aggressor Ghost camouflage it carried while assigned to the 64th FWS, was photographed at Holloman AFB on March 20, 1978. (Brian C. Rogers)

This T-38A, 62-3632, painted in the aggressor Grape scheme was one of the many 64th FWS T-38As which were transferred to the 479th TTW after being replaced in their aggressor role by F-5Es. (Ben Knowles)

Snake 68, T-38A 64-13168, was photographed at Holloman on March 20, 1978 after transferring from Nellis. (Brian C. Rogers)

Above: This white AT-38, 60-0594, of the 436th TFTS photographed in October 1978 carries the large HM (479th TTW Holloman) tail code and standard tail numbers 00594. (Don Logan)

Left: This AT-38, 60-0595, of the 434rd TFTS photographed on April 25, 1980, carries the small HM tail code, small 479th TTW insignia and small nonstandard tail numbers 60595. (Robert Burns via Jerry Geer)

A PICTORIAL HISTORY • 111

Above: Tail markings had not been standardized in October 1978 when this photo was taken. The 436th TFTS aircraft in the this photo, 65-10346, has the small HM and tail number, with black spades in the tail band. The 436th TFTS T-38 behind it has the large HM and tail number with a plain yellow band. (Don Logan)

Right: This AT-38, 61-0904, photographed in March 1978, has the large HM with small tail numbers. (Douglas Slowiak/Vortex Photo Graphic)

Below: 67-14842, with a black 42 nose number is seen here in October 1978 at Holloman in a special two tone blue camouflage. 67-14842 had been a 64th FWS aggressor and was painted with the Ghost camouflage scheme. (Don Logan)

This AT-38, 65-10370, an ex 64th FWS aggressor has "Check Six", meaning look behind you, in white letters in the tail band. (Brian C. Rogers)

This 436th TFTS AT-38, 62-3660, taxis for takeoff carrying a large tactical aircraft travel pod. The 436th TFTS insignia can be seen on the travel pod. (Brian C. Rogers)

This AT-38, 64-13188, another ex-64th FWS aggressor, was photographed at Holloman on October 14, 1980. (Ben Knowles)

Right and below: The 479th TTW developed a special three blue camouflage for the Wing's AT-38s. The colors of this camouflage are FS 35109, FS 35164, and FS 35450. 61-0812, in the AT-38 camouflage, was photographed on November 3, 1984 while assigned to the 433rd TFTS. It was marked as the 479th TTW Commander's (Flagship) aircraft including a multi-color tail band. (Craig Kaston)

65-10403, also marked as the 479th TTW Commander's aircraft, was photographed in June 1988. White edging has been added to the letters and numbers on the tail. (Ben Knowles)

67-14842 was marked as the 12th Air Force Flagship aircraft when photographed on October 11, 1991. (Keith Snyder)

63-8175 was marked as the 833rd Air Division Flagship when photographed in March 1988. (Ben Knowles)

61-0817 was photographed at Tinker AFB, Oklahoma during October 1989. It carries markings of the 433rd TFTS Flagship. (Jerry Geer)

During October 1990 60-0561 was photographed marked as the 436th TFTS Flagship. (Jerry Geer)

61-0820, assigned to the 434th TFTS, was carrying 479th TTW Commander's markings when photographed on August 27, 1988. (B.C. Rogers)

61-0876, assigned to the 435th TFTS, was photographed on March 24, 1984. (Brian C. Rogers)

65-10382, assigned to the 436th TFTS, was photographed on August 16, 1982. (Brian C. Rogers)

Left: The 479th TTW was inactivated on November 15, 1991. 68-8138 photographed at Luke AFB Arizona in June 1992 marked as the 49th FW Flagship and carrying an HO tail code. (Ben Knowles)

Below: After the inactivation of the 479th TTW, AT-38 operations were transferred to the 49th Operations Group with one remaining AT-38 squadron, the 433rd TFTS. 64-13298 is seen here on May 10, 1992, with an HO tail code, assigned to the 433rd TFTS. (Douglas Slowiak/ Vortex Photo Graphic)

When the 433rd TFTS was inactivated on July 8, 1992, its AT-38s were assigned to the 8th Fighter Squadron (FS). 64-13264 photographed on April 13, 1993, is marked as the 8th FS "Black Sheep" Flagship. (Brian C. Rogers)

ACCELERATED CO-PILOT ENRICHMENT (ACE) PROGRAM

As a result of the wind-down of UPT requirements resulting from the end of U.S. involvement in Vietnam, the USAF had excess T-38s in inventory. This excess in ATC's inventory led to the assignment of numerous Talons to other USAF commands for a variety of training programs. In 1974 the Talon joined SAC, employed as an economical alternative to training copilots to command standard and thus avoiding extra utilization of the force's bomber and tanker fleet. In the mid-1970s, SAC Squadrons had an excess of copilots. The T-38 gave the SAC organizations a less expensive means of "seasoning" the copilots, giving them hands-on flying experience and allowing them to improve their decision making skills. ACE (Accelerated Copilot Enrichment) program usually has six T-38s or T-37s on hand. ACE trainers maintained and staffed by ATC personnel were used by SAC bomber and tanker units, and the 9th SRW, at Beale AFB, California. The ACE program was redesignated Companion Trainer Program (CTP) after SAC was inactivated.

BASE	UNIT	SUPPORTING ATC BASE	UNIT
Dyess AFB, TX	96th BW	Randolph AFB	12th FTW
Ellsworth AFB, TX	28th BW	Reese AFB	64th FTW
Grand Forks AFB, N.D.	319th BW	Reese AFB	64th FTW
Malmstrom AFB, MT	301st ARW	Reese AFB	64th FTW
March AFB, CA	22nd BW/ARW	Williams AFB	82nd FTW
Minot AFB, N.D.	5th BW	Reese AFB	64th FTW
Offutt AFB, NE	55th SRW	Randolph AFB	12th FTW

62-3628 was an ACE aircraft assigned to the 64th FTW at Reese AFB Texas. As seen in this June 1977 photo, the ACE T-38s of the 64th FTW had the speed brakes painted as an ACE of Spades. (Ray Leader)

62-3646 photographed on October 1, 1976, sits at the end of a line of ACE T-38s at Grand Forks AFB North Dakota. (Douglas Slowiak/Vortex Photo Graphic)

COMPANION TRAINER PROGRAM (CTP)

T-38s (many from the ACE program) are being used by non-fighter Air Combat Command units. ACC wings with bomber or reconnaissance aircraft are part of the companion trainer program. The 49th FW at Holloman AFB flying F-117s is the only fighter unit using T-38 companion trainers. The T-38 in these units are, for the most part, painted in the same paint schemes as the Wing's primary aircraft. Several AMC Tanker units also operated T-38As during 1993 before converting to C-12Fs for companion trainers. One KC-135 tanker squadron, the 22nd Air Refueling Squadron of the 366th Wing at Mountain Home AFB, and one KC-10 tanker squadron, the 344th Air Refueling Squadron of the 4th Wing at Seymour-Johnson AFB, fly the T-38 as a companion trainer.

COMPANION TRAINERS

TAIL CODE	WING	SQUADRON	BASE	PRIMARY AIRCRAFT
BB	9 RW	1st RS	Beale AFB	U-2R/KC-135Q
DY	7/96 WG	9th BS	Dyess AFB	B-1B/KC-135
EL	28 BW	37th BS	Ellsworth AFB	B-1B
GF	319 BW/BG*	46th BS	Grand Forks AFB	B-1B
HO	49 FW	7th FS	Holloman AFB	F-117
LA	2 BW	20th BS	Barksdale AFB	B-52H
MO	366 WG	22nd ARS	Mountain Home AFB	KC-135R
MT	5 BW	23rd BS	Minot AFB	B-52H
OF	55 WG	24th RS	Offutt AFB	RC/EC-135
OZ	384 BW/BG*	28th BS	McConnell AFB	B-1B
SJ	4 WG	344th ARS	Seymour-Johnson	KC-10
WM	509 BW	393rd BS	Whiteman AFB	B-2
–	22 ARW*		March AFB	KC-10
–	43 ARW*		Malmstrom AFB	KC-135

* These Wings are no longer flying T-38s as companion trainers. The 22nd ARW is now based at McConnell flying KC-135R/Ts with C-12s as companion trainers. The 43rd ARW is still at Malmstrom but is flying C-12s as companion trainers. The 319th is now an Air Refueling Wing, still at Grand Forks flying the KC-135R as its primary aircraft with C-12 as companion trainers. The 384th has been inactivated.

Below: The 49th Fighter Wing uses the T-38 as a companion trainer. The gloss black T-38s are assigned to the 7th Fighter Squadron and are used to support F-117 pilot training. Since no two seat F-117s have been built, and as part of F-117 pilot training, an instructor pilot flying a T-38 is required to fly alongside the student pilot in the F-117 during the initial phase of training. (USAF)

BEALE AFB

9th WG

Right and below: After SAC was inactivated and the 9th Strategic Reconnaissance Wing transferred to ACC, the unit adopted the tail code BB (Beale Bandits). 64-13304, seen here in March 1993, carries the name "Sierra Special" on the nose. (Douglas Slowiak/Vortex Photo Graphic)

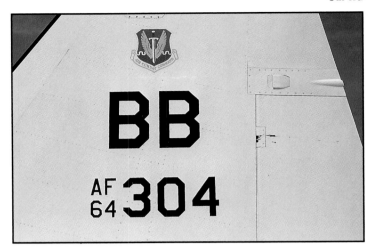

64-13301 is seen at Luke AFB on April 12, 1993, prior to being painted flat black with red markings. Many of the companion trainers were repainted at Luke AFB. 64-13301 is carrying the name "Wild Thing" on the nose. (Don Logan)

Left and below: 64-13301 is seen here at Langley AFB in August 1993 after being repainted to the 9th Wing's colors of black and red. (Don Logan)

64-13271, the 9th SRW flagship from as far back as 1979, is seen in its black and red on August 7, 1994. (Brian C. Rogers)

While belonging to the 96th Wing, KC-135 pilots flew T-38s as companion trainers. 64-13179, photographed on July 26, 1993, was one of three T-38s which carried 917th Air Refueling Squadron tail band. (Don Logan)

B-1B pilots of the 96th Wing flew T-38s as companion trainers. The 64-13206 337th Bomb Squadron (BS) was responsible for T-38 operation for both the 337th BS and 338th CCTS. 64-13206 is seen here during July 1993 with the 337th BS tail stripe. (Don Logan, Keith Snyder)

67-14952, with a 337th BS tail stripe painted gunship gray (FS 36118), was photographed on July 26, 1993. The B-1Bs assigned to the 96th Wing were also painted gunship gray. (Don Logan)

In early 1994 the 96th Wing was inactivated and replaced with the 7th Wing which had been at Carswell AFB in Forth Worth, Texas. Under the 7th Wing, the 9th BS replaced the 337th as the operator of T-38s. 68-8107 is seen here on July 27, 1994, with the 9th BS tail marking of a white bat on a black stripe. (Don Logan)

ELLSWORTH AFB   **28 BW**

EL tail coded 62-3634, assigned to the 37th Bombardment Squadron, 28th Bombardment Wing, seen here at Ellsworth AFB, South Dakota on May 11, 1993, before receiving its gunship gray paint. (John Cook)

Photographed in its gunship gray paint on June 1, 1994, 66-4322 of the 37th BS awaits the flight crew for a training flight. (Don Logan)

GRANDFORKS AFB 319th BW/BG

Photographed in October 1993, GF tail coded "Musky Two", serial number 64-13268, assigned to the 46th BS, 319th BW at Grand Forks AFB, North Dakota was photographed on the McConnell AFB flight line. A close-up of the tail shows it has been marked in error as 67-3268, which is not a T-38 serial number. (Tail – John Cook, overall view – Jim Geer)

"Musky Four", 64-13175 from Grand Forks AFB photographed at McConnell AFB on March 30, 1994, is still in its gloss white paint without the GF tail code. (Don Logan)

HOLLOMAN AFB 49th FW

HO coded 68-8150, seen here on September 8, 1994, in gloss black paint, is assigned to the 7th Fighter Squadron of the 49th Fighter Wing. Initially, even though the 49th's T-38s carry tail codes, their tail numbers were presented in the ATC format of five 12 inch high digits than the standard tactical serial number format of "AF", year, and the last three digits. (Brian C. Rogers)

86-8177, prior to receiving the F-117 tail stripe, photographed on landing approach. (Ben Knowles)

68-8177, at Langley AFB on July 28, 1994. (Brain C. Rogers)

As seen here on 68-8177 photographed at Langley on December 29, 1994, the 49th FW's T-38s have adopted the standard tactical serial number format. (Brian C. Rogers)

BARKSDALE AFB **2nd BW**

The T-38s of the 2nd Bomb Wing at Barksdale AFB Louisiana are operated for the Wing by 20th Bomb Squadron (BS). 67-14833 is seen here in late December 1994 with a the 20th BS blue tail stripe and LA tail code. (Terry Somerville)

66-4383 is also assigned to the 2nd Bomb Wing at Barksdale. (Terry Somerville)

126 • NORTHROP'S T-38 TALON

64-13197 had just arrived from Seymour Johnson when this photo was taken in late December 1994. SJ tail code had been painted out, and the LA tail code had not yet been applied. (Terry Somerville)

Above and left: This 2nd BW aircraft, 64-13206 previously at Dyess AFB, had not yet had the blue tail stripe painted when photographed in late December 1994. (Terry Somerville)

MOUNTAIN HOME AFB

366th WG

The 22nd Air Refueling Squadron of the 366th Wing at Mountain Home AFB, Idaho, the only ACC unit flying KC-135R tankers, uses T-38s as companion trainers. This 366th Wing white T-38, 62-3652 was photographed at Langley AFB Virginia on September 16, 1994. A full color 22nd ARS insignia is on the left side of the nose, with a full color 366th Wing insignia on the right side of the nose. (Brian C. Rogers)

62-3640, in gunship gray (FS 36118) was photographed on October 11, 1993 at Tinker AFB. (Randy Walker)

MINOT AFB **5th BW**

65-10419 assigned to the 23rd BS, 5th BW at Minot AFB, North Dakota was photographed at McConnell AFB on February 6, 1994. (John Cook)

This Minot jet, 62-3674, prepares to taxi for a return flight to Minot on March 20, 1993. (Norris Graser)

A PICTORIAL HISTORY • 129

OFFUTT AFB — 55th WG

Right and below: The 24th Reconnaissance Squadron of the 55th Wing stationed at Offutt AFB, Nebraska operates T-38s. 66-4383 in gloss white and 67-14923 in gray (FS 36173) are parked on the ramp at Offutt AFB during June 1993. (Don Logan)

67-14939 is seen in June 1993 being towed to the 55th Wing maintenance hangar. (Don Logan)

The 384th's T-38s sit on the McConnell AFB, Kansas flight line under threatening clouds. (Don Logan)

66-4388 in white, assigned to the 28th BS of the 384th BW, was photographed in front of the ACE hangar on May 13, 1993. The 384th Wing tail code, OZ, is a reference to the Land of OZ contained in the book "The Wizard of OZ." (Don Logan)

66-4388 photographed on September 15, 1993, in its recently applied coat of gunship gray paint. (Don Logan)

SEYMOUR-JOHNSON AFB — 4th WG

This aircraft, 67-14826, photographed at Luke AFB, Arizona had recently been refinished in FS 36173, the same paint color being applied to the AMC tanker/transport fleet. The SJ tail code indicates the aircraft is assigned to the 344 Air Refueling Squadron of the 4th Wing at Seymour-Johnson AFB, North Carolina. Note that the tail number has been incorrectly applied as 64-4826. (Don Logan)

64-13197 photographed on September 15, 1993, in its recently applied coat of AMC gray paint. (Ben Knowles)

WHITEMAN AFB — 509 BW

The B-2 Wing, the 509 BW based at Whiteman AFB, Missouri, uses T-38s as its companion trainer. 68-8179, operated by the 393rd BS, was photographed on March 28, 1994, during a visit to Langley AFB, Virginia. (Brian C. Rogers)

The 22nd Air Refueling Wing at March AFB, California operated T-38s as ACE trainers until October 1993. This T-38, 63-8225, the last March T-38, was photographed on the receiving line at AMARC in October 1993. (Douglas Slowiak/Vortex Photo Graphics)

THUNDERBIRDS

The year 1972 saw the last of the T-38s delivered to the USAF. Two years later the Talon was chosen as the new aircraft for the 3600th Aerial Demonstration Squadron – the "U.S. Air Force Thunderbirds." Using the T-38 trainer for the Thunderbirds was a radical departure from previous practice of use of a front line fighter as the team aircraft. The previous Thunderbird aircraft had started with the F-84 in 1953, and included the F-100 and F-105 in the early 1960s, and finally the F-4 Phantom II in 1968. After six years with the F-4, the fuel economy conscious attitudes of the 1970s made use of a more fuel efficient aircraft good for public opinion, and would also be less expensive to operate. Many European flight demonstration teams had also transitioned to trainers so the USAF was not alone in this change to a more economical aircraft. Even with the economic factors considered, it was not an easy decision to leave the tradition of first line fighters and use a trainer as the Thunderbirds' aircraft. It had long been stated that the USAF T-birds pilots were line pilots flying the same type of aircraft as the operational squadrons in Tactical Air Command. Now the aircraft was going to be a trainer, but at least it was a supersonic high performance aircraft.

For the Thunderbirds role, the Talon received an entirely new paint scheme. The Talons streamlined airframe was not suited to the Thunderbird motif applied to the underside contours of previous aircraft types. Instead, the Talon had gracefully curved stripes applied to the fuselage and vertical tail. Under the wings a double arrowhead sweeping forward from the wingtips to nose was applied. This achieved a good contrast between the upper and lower surfaces when seen in plan view or during the fast rolls for which the T-38 was well known.

The T-Birds operated the T-38 from 1974 until 1982. During this time the T-Birds had three fatal accidents. The first occurred on 9 May 1981. The opposing solo pilot was killed when his T-38 crashed inverted at Hill AFB, Utah. The aircraft crashed outside the air base apparently with both engines stopped. On September 8, 1981 a second T-bird Talon crashed at Cleveland Airport, Ohio. Departing after completing the show, team leader Lt.Col. David Smith had a multiple birdstrike when he ran into a flock of seagulls. Smith and his groundcrew passenger ejected at low altitude but Smith's parachute failed to open in time. His ground crewman survived having suffered only minor injuries.

The final fatal accident occurred during practice at Indian Springs air field, north west of Nellis on the morning of 10 January 1982. The four aircraft formation (Numbers 1 through 4) struck the ground at the bottom of a line abreast loop. All four pilots were killed. Examination of the crash site indicated that the Thunderbird leader, Major Lowry impacted first, followed by Number 4, with Numbers 2 and 3 hitting last. As the flight of four neared the bottom of the loop, the stabilizer of the lead aircraft (No.1) apparently jammed, preventing the aircraft from recovering at the bottom of the loop. The other three pilots flying standard formation procedures were only watching the lead aircraft and did not see their closeness to the ground until it was too late. The 1982 show season was canceled before it had even begun. The T-38 was replaced by the F-16 for the 1983 show season.

THUNDERBIRD T-38s

USAF SERIAL NUMBER	THUNDERBIRD NUMBER
68-8100	8
68-8106	5/2
68-8131	1/8
68-8137	7
68-8156	6/3
68-8174	2/7
68-8175	7/4
68-8176	9/3/6
68-8177	4/8/5
68-8182	9
68-8183	1
68-8184	3/5

The primary fourship and the two solos join together to form the Thunderbird Delta. From this formation the Delta Loop and Delta Roll are performed. (USAF via Steve Haskin)

The Thunderbird Diamond formation as seen from below. (USAF)

The two solos join up with lead solo (#5) inverted. (USAF via Steve Haskin)

A PICTORIAL HISTORY • 135

Above and below: During late 1973 and early 1974 the Thunderbirds practiced with T-38s displaying partial "practice" markings 68-8174 – #2, and 68-8156 – #6 are seen on the Nellis flight line in March 1974. (Don Logan)

Right: The Thunderbird on the far side of this formation carries non-standard tail markings. (USAF via Steve Haskin)

The Thunderbird markings included the flight position number (1, 2, 3, 4, 5, or 6) on the tail. Additional training and support aircraft were numbered 7, 8, and 9. (Don Logan)

The flags of the countries in which the team had flown a show are displayed on the left side of the aircraft below the rear canopy. (Don Logan)

The Thunderbird Team insignia was displayed on the right side of the fuselage below the rear canopy. (Don Logan)

The Thunderbird Team seen here in dress blue uniforms assembled with seven of the aircraft on January 27, 1975, for a change of command ceremony. (Don Logan)

A PICTORIAL HISTORY • 137

During the Bi-Centennial year (1976) the aircraft numbers were shifted to the intake, and the space on the tail was filled with the Bi-Centennial emblem. (Don Logan)

Aircraft 1, 9, 7, 6, and 5 line up on the runway at Nellis for a practice mission. (Don Logan)

The Thunderbirds, flying practice aircraft, perform a show at Nellis on February 14, 1975. (Don Logan)

AVIATION INDUSTRY T-38s

The T-38 has been used by a number of civilian aerospace companies as both instrumented chase aircraft and avionics test bed aircraft. Lockheed-Georgia used four T-38As on loan from the USAF as chase aircraft on both the C-141 and C-5 test programs.

Serial Number	Program	Dates
59-1600	C-5A	June 3, 1968 - July 2, 1974
59-1601	C-5A	June 3, 1968 - June 28, 1971
59-1603	C-141	February 12, 1964 - July 7, 1966
59-1604	C-141	February 12, 1964 - July 7, 1966

General Dynamics-Fort Worth used 65-10402 on loan from the USAF in the F-16 program starting in September of 1972 and later transferred to 377th TES at McClellan AFB.

PRIVATELY OWNED T-38s

Chuck Thornton of The Thornton Company in Van Nuys, California restored two T-38s and they are presently used for civilian flight test. A third T-38 is under restoration by Thornton. In addition, his company's T-38s have been used by Hollywood in movies such as "Dragnet" and "Hot Shots."

N638TC is equipped with flight test instrumentation which allows it to fly as a flight test chase aircraft.

Chuck Thornton's first T-38 N538TC was sold to The Boeing Company and has been used in their flight test programs replacing an F-86 and a T-33. Boeing's T-38, N38FT, has been used in both military and civilian flight test programs, including the program which installed probe and drogue wing tip refueling pods on French Air Force C-135FRs, and is presently being used as part of the 777 flight test program.

61-00862 Northrop # 5228 N538TC
Restored by Chuck Thornton as N538TC and later sold to Boeing renumbered as N38FT.

63-08171 Northrop # 5518 N638TC
Restored by Chuck Thornton as N638TC used by the Thornton Corporation for flight test support.

65-10462 Northrop # 5881
Presently under restoration at Van Nuys Airport, Van Nuys, California.

59-1604, seen here in AFSC markings, was used in the C-141 test program. (Ben Knowles)

T-38 61-0908, seen here in October 1975 flying chase on a B-52D 55-0087 during B-52D Pacer Plank (Wing Modification) testing. Pacer Plank mod was accomplished at Boeing's Wichita, Kansas facility. Note the external wing tip fuel tanks have been removed, and as a result the wings are bowed up above the fuselage. (Boeing)

Boeing uses a T-38 which it owns as an instrumented chase aircraft. It's seen here escorting the 777 on its first flight. (Boeing)

61-0862, rebuilt by Chuck Thornton and given the registration number N538TC, is seen here at Mojave airport in California. It still carries the black paint applied for its part in the movie "Hot Shots." (Don Logan Collection)

Above and below: N538TC was purchased from Chuck Thornton by Boeing and was repainted and re-registered as N38FT. N38FT is seen in these two photos at Boeing's Wichita facility departing on a test mission. (Boeing)

Right: Chuck Thornton's second T-38, N638TC (ex 63-8171), is seen here flying chase with a Rockwell B-1B over the Mojave desert. (Rockwell International)

Below: N638TC is seen here on Cessna's Wichita flight line. The T-38 was under contract to Cessna to support flight test of the new Citation X. (Don Logan)

In addition to special communication and instrumentation systems, Thornton's T-38 have an F-5 boat tail which adds a drag chute capability to his T-38s. (Don Logan)

AIRCRAFT DESCRIPTION/SYSTEMS

The T-38A aircraft is a two-place, twin-turbojet super sonic trainer. The fuselage is an area-rule (coke bottle) shape, with moderately swept-back wings and empennage. The aircraft is equipped with an all movable horizontal tail. The tricycle landing gear has a steerable nosewheel. The overall dimensions of the aircraft are:

Length	46 ft. 4 in.
Wingspan	25 ft. 3 in.
Height	12 ft. 11 in.
Tread	10 ft. 9 in.
Wheelbase	19 ft. 5 in.

The average gross weight of the aircraft fully fueled and including two aircrew is 12,000 pounds.

The high thrust-to-weight ratio of its twin J-85-5 engines in a lightweight airframe gives the Talon a top speed of Mach 1.24 in level flight, a sea-level rate of climb of 30,000 feet per minute, and a service ceiling of 54,000 feet. The aircraft reaches Mach 1.55 in dives. At normal gross weight of 11,550 pounds, it requires only 2,350 feet for takeoff.

A fully-loaded T-38A with landing gear and flaps down is capable of takeoff, go-around and landing with only one engine operating. On one engine the aircraft has a sea-level rate of climb of 6,800 feet per minute and a service ceiling of 45,000 feet, and can reach Mach .95 in straight and level flight.

Structure of the T-38A Talon is made of riveted aluminum alloy, semi-monocoque with stressed skin; however, considerable use also was made of adhesive-bonded aluminum honeycomb and chemical-milled structural parts. The airframe is designed for a load limit of 7.33 g, with an ultimate load capability of at least 1.5 times this value for added safety.

The thin dry wing was built as a single unit from tip to tip, with multiple spars and thick aluminum alloy skins. The trailing-edge flaps, ailerons, detachable wingtips and various sections of the wing are honeycomb-stiffened structures. The all-movable ("flying") horizontal stabilizer and rudder are also stiffened with honeycomb. The aft removable section of the fuselage (called the boat tail) surrounding the engine tailpipes and supporting the horizontal stabilizer is made up of a combination of different metals, including steel and titanium.

AIRCRAFT SYSTEMS

ENGINES

The aircraft is powered by two J85-GE-5 series, eight-stage, axial-flow, turbojet engines. Sea level, standard day, static thrust for an uninstalled engine is 2680 pounds (approximately 2050 pounds installed) at MIL power and 3850 pounds (approximately 2900 pounds installed) at full MAX power. Air enters through the variable inlet guide vanes which direct the flow of air into the compressor. Automatic positioning of the inlet guide vanes and air bleed valves assists in regulating compressor airflow and maintains compressor stall-free operation. Two turbine wheels aft of the combustion section are on the same shaft as the compressor rotor stages. The exhaust gases are discharged through a variable area exhaust nozzle. An exhaust gas temperature sensing system varies the nozzle area to maintain exhaust gas temperature within limits at both MIL and MAX range throttle positions. Each engine has a main fuel control system and an afterburner fuel control system. The main fuel control system consists primarily of a two-stage engine driven pump, a main fuel control, and an overspeed governor.

Afterburner operation is initiated by advancing throttle from the MIL detente into the MAX range. Thrust is variable within MAX range. The total rate of fuel flow at full MAX position for each engine at sea level on a standard day is approximately 7300 pounds per hour with the aircraft at rest and 11,400 pounds per hour at Mach 1.

Throttles

Two pairs of throttles, one pair in the front cockpit and one pair in the rear cockpit, have OFF, IDLE, MIL and MAX positions, with the speed brake switch and the microphone button on the right throttle. The throttles in the front cockpit are equipped with fingerlifts which must be raised before the throttles in either cockpit can be retarded past IDLE to OFF. Throttle friction, which increases/decreases the force required to move the throttles, is ground adjustable only. The throttles, when placed at OFF, mechanically shut off fuel to the engine at the main fuel control and electrically shut off fuel to the engine at the fuel shutoff valves.

Engine Start And Ignition System

Engine starting requires a low pressure air supply to rotate the engine's compressors, DC power to energize the ignition holding relay, and AC power for ignitor firing. An engine start pushbutton for each engine is located on the left sub-panel of each cockpit. For ground starts only, momentarily pushing a start button positions the diverter valve to direct low pressure air to the selected engine for compressor rotation, and arms the ignition circuit for approximately 30 seconds. Moving the throttle to IDLE on the ground or in flight energizes the ignition system and supplies starting fuel flow to the engine. Moving the throttles to MAX range in flight or on the ground energizes the main and afterburner ignitors to light off the afterburners. AC power from a battery operated static invertor is used for ground start (one engine) or air starts (either engine). For battery start, the right engine must be started first, since the static invertor supplies AC power for the right engine instruments only during the start cycle.

Engine Instruments

A full complement of engine instruments including engine tachometers, exhaust gas temperature (EGT) indicators, nozzle position indicators, fuel flow indicators, and hydraulic pressure indicators is provided in each cockpit.

Oil System

Each engine has an independent integral oil supply and lubrication system. The reservoir has a normal oil capacity of four quarts and an air expansion space of one quart. Heat from the engine oil is dissipated through a fuel-oil cooler.

Fire Warning And Detection System

A fire warning and detection system is provided to give the both crewmembers warning of a fire or overheat condition in either engine bay. The system includes a temperature sensing loop in the forward and aft sections of each engine bay, and a left engine and right engine fire warning light in each cockpit. Operation of the warning and detection system in each engine bay is independent of the other, except when testing the system from the cockpit.

FUEL SYSTEM

The aircraft has an independent fuel supply system for each engine, interconnected by a crossfeed valve. The left and right system fuel cells are in the fuselage. The left engine is supplied by the forward fuselage cell and the forward and aft dorsal cells; the right engine, by the center and aft fuselage cells. An electric fuel boost pump in each system supplies fuel under pressure to the engine driven fuel pump during normal operation. The fuel boost pumps are required for inverted flight. Without the aid of the boost pump, each engine can be supplied with fuel by gravity flow from its respective system. Normally, sufficient fuel will flow by gravity to maintain MAX power from sea level up to approximately 25,000 feet; however, gravity flow is guaranteed only to 6,000 feet, and flameouts have occurred as low as 15,000 feet. Through crossfeed operation, both systems may supply fuel to either engine with or without boost pump pressure.

PRESSURE AIR SYSTEM

Air taken from the engines' eighth stage compressor is used for hydraulic reservoir and cabin pressurization, air conditioning systems, canopy defogging, engine anti-icing, canopy seal inflation, and for the anti-G suit system.

AIRFRAME-MOUNTED GEARBOX

An airframe-mounted gearbox for each engine operates a hydraulic pump and an AC generator. A shift mechanism keeps AC generator output between 320 and 480 cycles per second. Automatic gearbox shift occurs in the 65% to 70% RPM range.

ELECTRICAL SYSTEMS

Two alternating current systems and one direct current system supply electrical power to the aircraft. The 115/200-volt AC power supply systems consist of two identical engine-driven AC generating systems, one mounted on each airframe mounted gearbox, and an external power receptacle. The DC power supply system consists of a DC bus powered either by a 24-volt, 5-ampere-hour battery or two 28-volt DC transformer-rectifiers.

HYDRAULIC SYSTEMS

Two aircraft hydraulic power supply systems, a 3000-psi utility hydraulic system powered by the left engine and a 3000-psi flight control hydraulic system powered by the right engine. The two systems are completely separate, and no fluid flow can occur between them. Separate pressure indicators and caution lights are provided for each system.

FLIGHT CONTROL SYSTEM

The T-38 uses a conventional flight control system with artificial feel is provided. Each primary flight control surface is actuated by two independently powered hydraulic cylinders, one actuated by the utility hydraulic system and the other by the flight control hydraulic system. The ailerons and horizontal tail are trimmed by electrical actuators, which change the relationship of "feel" springs to the control sticks.

WING FLAP SYSTEM

The flaps, located on the trailing edge of the wing next to the fuselage, are electrically controlled by a flap lever on the aft face of the throttle quadrant. Two AC electric motors operate the flaps through gear reduction units. The flaps are interconnected by a rotary flexible shaft. If one flap motor fails, both flaps are actuated through the rotary shaft. Full flap extension or retraction takes from 10 to 17 seconds. Flaps are mechanically interconnected to the horizontal tail. The flap-to-horizontal tail interconnecting linkage moves the tail trailing edge down as the flaps are lowered, and moves the trailing edge up as the flaps are raised. This coordinated movement reduces the need for manual input of aircraft trim changes by the pilots as the flaps are lowered or raised.

SPEED BRAKE SYSTEM

An electrically controlled, hydraulically actuated dual speed brakes are located on the lower surface of the fuselage center section just forward of the main gear wheel wells. The speed brakes are variable position type, opening fully in approximately four seconds and closing in approximately three seconds.

LANDING GEAR SYSTEM

Extension and retraction of the landing gear and gear doors are powered by the utility hydraulic system and electrically controlled by the landing gear levers. Landing gear extension or retraction normally takes approximately six seconds. The normal extension sequence is doors open, gear extends, doors close. The retraction sequence is doors open, gear retracts, doors close.

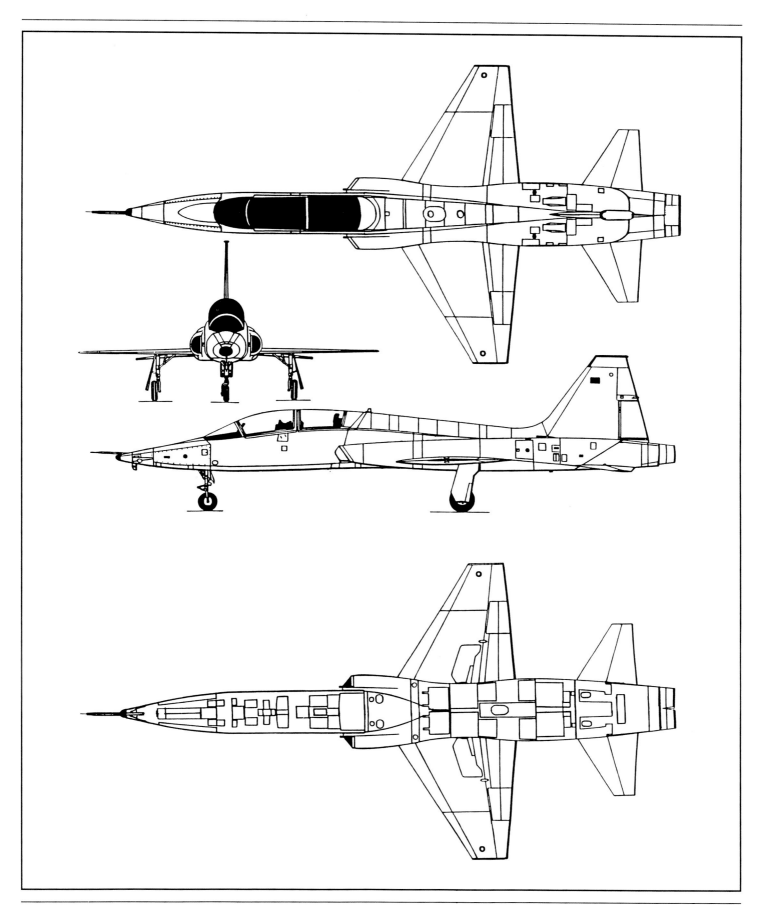

FRONT COCKPIT INSTRUMENT PANEL

1. VERTICAL SITUATION DISPLAY
2. HORIZONTAL SITUATION DISPLAY
3. ALTIMETER
4. VERTICAL SPEED INDICATOR
5. TACHOMETERS
6. TEMPERATURE EXHAUST INDICATORS
7. NOZZLE POSITION INDICATORS
8. RIGHT FIRE WARNING LIGHT
9. OIL PRESSURE INDICATORS
10. FUEL QUANTITY INDICATORS
11. FUEL CONTROL PANEL
12. CANOPY JETTISON "T" HANDLE
13. ELECTRICAL CONTROL PANEL
14. TORSO AIR OUTLET
15. LIQUID OXYGEN QUANTITY INDICATOR
16. OXYGEN REGULATOR
17. WARNING LIGHTS PANEL
18. TRANSPONDER SET CONTROL
19. CODER GROUP CONTROL PANEL
20. COCKPIT LIGHTING PANEL
21. FUEL FLOW INDICATORS
22. AIR CONDITIONING AND PRESSURIZATION CONTROL PANEL
23. HYDRAULIC PRESSURE INDICATOR
24. UTILITY HYDRAULIC PRESSURE INDICATOR
25. COMMUNICATION PANEL
26. TACAN (TACTICAL AIR NAVIGATION PANEL)
27. CIRCUIT BREAKERS
28. ILS (INSTRUMENT LANDING SYSTEM) PANEL
29. INTERCOM SELECTOR PANEL
30. INTERCOMMUNICATIONS AND RADIO TRANSFER PANEL
31. MICROPHONE BUTTON
32. SPEED BRAKE SWITCH
33. STABILITY AUGMENTER PANEL
34. TAKEOFF TRIM PANEL
35. FLAP CONTROL
36. ENGINE THROTTLES
37. STOWAGE BOX
38. COMPASS FAST SLAVE SWITCH
39. ENGINE START PANEL
40. ALTERNATE LANDING GEAR RELEASE
41. LANDING AND TAXI LIGHT SWITCH
42. FLAP POSITION INDICATOR
43. LANDING GEAR CONTROL PANEL
44. LEFT FIRE WARNING LIGHT
45. CABIN ALTIMETER
46. ACCELEROMETER
47. STANDBY COMPASS
48. NAVIGATION MODE SELECTOR PANEL
49. AIRCRAFT CLOCK
50. MACH/AIRSPEED INDICATOR
51. MARKER BEACON LIGHT

REAR COCKPIT INSTRUMENT PANEL

1. VERTICAL SITUATION DISPLAY
2. HORIZONTAL SITUATION DISPLAY
3. MASTER CAUTION LIGHT
4. ALTIMETER
5. CANOPY LOCK LIGHT
6. VERTICAL SPEED INDICATOR
7. TACHOMETERS
8. TEMPERATURE EXHAUST INDICATORS
9. NOZZLE POSITION INDICATORS
10. RIGHT FIRE WARNING LIGHT
11. OIL PRESSURE INDICATORS
12. FUEL QUANTITY INDICATORS
13. FUEL CONTROL PANEL
14. CANOPY JETTISON "T" HANDLE
15. TORSO AIR OUTLET
16. LIQUID OXYGEN QUANTITY INDICATOR
17. WARNING LIGHTS PANEL
18. OXYGEN REGULATOR
19. FUEL FLOW INDICATORS
20. FLIGHT CONTROL HYDRAULIC PRESSURE INDICATOR
21. UTILITY HYDRAULIC PRESSURE INDICATOR
22. COMMUNICATION PANEL
23. TACAN (TACTICAL AIR NAVIGATION PANEL)
24. ILS (INSTRUMENT LANDING SYSTEM) PANEL
25. INTERCOM SELECTOR PANEL
26. COMMUNICATION AND NAVIGATION OVERRIDE PANEL
27. AIR START SWITCH
28. FLAP POSITION INDICATOR
29. LANDING GEAR CONTROL PANEL
30. LEFT FIRE WARNING LIGHT
31. ACCELEROMETER
32. NAVIGATION MODE SELECTOR PANEL
33. AIRCRAFT CLOCK
34. MACH/AIRSPEED INDICATOR
35. MARKER BEACON LIGHT

EJECTION SYSTEM

The ejection system consists of an ejection seat with a man-seat separator, an automatic opening safety belt with a one-second delay initiator, and a parachute with an aneroid and one-second timer function and a "One and Zero" escape capability.

Approximately one second after ejection from the aircraft, the safety belt initiator fires, opening the safety belt, and actuates the man-seat separator, forcing the crewmember from the seat. The parachute arming lanyard arms the parachute as the crewmember separates from the seat. When freefalling from high altitude, a timer is activated at a preset altitude to give a full chute at approximately 14,000 feet pressure altitude.

To provide a low altitude "One and Zero" escape capability, a zero delay lanyard is hooked to the parachute ripcord handle. After ejection from the aircraft, the safety belt opens and the crewmember is separated from the seat. The zero delay lanyard pulls the parachute ripcord handle, providing immediate deployment of the parachute.

OXYGEN SYSTEM

The T-38 uses a liquid oxygen system to supply breathing oxygen to crewmembers. The oxygen regulators (automatic diluter demand) control the flow and pressure of the oxygen and distribute it in the proper proportions to the masks. An oxygen regulator is on the right console of each cockpit.

ANTI-G SUIT SYSTEM

Air pressure from the air-conditioning system is used to inflate the anti-G suit in each cockpit to offset the effects of high G load factor.

COMMUNICATION AND NAVIGATION EQUIPMENT

The communication and navigation equipment on the T-38 is made up of a UHF Command Radio (AN/ARC-34), an Intercom for voice communications between both cockpits and with the ground crew, a TACAN (AN/ARN-65), an ILS receiver (AN/ARN-58), and an IFF/SIF or AIMS transponder for air traffic control uses.

AT-38B SPECIAL EQUIPMENT

A number of standard T-38As were modified to AT-38B Lead-In-Fighters (LIF) by the addition of a centerline pylon with weapons carriage capability and a noncomputing gunsight. Some aircraft have the MXU-553 flight loads recorder installed. Both cockpits contain additional controls for the armament system.

Each cockpit's control stick grip was replaced with a standard fighter type stick grip containing, in addition to the standard flight control trim switch and nosewheel steering button, a bomb-rocket button and a trigger. In the front cockpit, the bomb-rocket button and trigger actuate the KB-26A camera as well as release armament. A pylon is mounted on the centerline of the aircraft and is equipped with an MA-4 bomb rack capable of handling stores up to 1000 pounds with 14" suspension lugs. The pylon is bolted to the airplane and is not jettisonable. Stores are released from the pylon either through the armament release system or the emergency jettison system; in either case the bomb rack hooks are opened and the store is allowed to fall free. The pylon does not have the capability of forced weapon ejection.

The aircraft has the mission capability of delivering a variety of non-nuclear training munitions in order to introduce and teach the basic delivery techniques to new pilots. When carrying a SUU-20 or AF-B37K-1 bomb rack, level bomb delivery and dive bomb delivery of 0 to 45 degrees can be practiced using BDU-33 series bombs. The same cockpit armament controls are used to release bombs from either the SUU-20 or the AF-B37K-1. In addition, the SUU-20 has the capability of launching four 2.75 inch folding fin aircraft rockets (FFARs). Strafing can be practiced using the SUU-11A/A or SUU-11A/B 7.62 mm minigun pod.

SUU-20 Bomb-Rocket Dispenser

Three types of SUU-20 series bomb-rocket dispensers (SUU-20/A(M), SUU-20A/A, or SUU-20B/A) can be carried by the AT-38. The dispensers are capable of carrying and delivering six BDU-33 series practice bombs and four 2.75 inch FFAR. All three dispensers are similar in design and can be carried on the centerline pylon. The center section of the SUU-20 is recessed for carrying six practice bombs (pairs in tandem), and on each side of the recessed area are rocket tubes (two on each side) for carrying the rockets. Each bomb is safetied in position with a red flagged ejector safety pin which must be removed before flight. The pin mechanically locks the bomb in the dispenser. The four rocket tubes use a detente latch to hold the rocket in the tube.

AF/B37K-1 Bomb Container

The AF-B37K-1 bomb container is a practice bomb carrying device which can carry and release four BDU-33 practice bombs, and can be mounted on the centerline pylon. The bomb container is constructed of two steel plate U-shaped channels that form an inner and outer shell from which four bomb racks are suspended (pairs in tandem). Each rack can hold one practice bomb.

BDU-33 Series Practice Bombs

BDU-33 practice bombs are carried by both the SUU-20 and the AD-B37K-1. The bombs have a cast iron body with an attached fin assembly. The bombs produce white smoke on contact with the ground, which allows the impact to be visually scored.

2.75-Inch Folding Fin Aircraft Rocket (FFAR)

The 2.75-inch folding fin rocket is used to deliver a variety of warheads against ground targets. The complete round consists of a motor, warhead, and fuse. The FFAR when loaded on the AT-38 uses inert practice warheads.

SUU-11A/A -11B/A Gun Pod

An SUU-11A/A, -11B/A, 7.62-millimeter "Minigun" pod houses a six-barrel, electrically operated GAU-2B/A "Gatling" type gun which provides rates of fire up to 6000 rounds per minute (SPM) at sustained fire. The SUU-11B/A gun pod has a rate

SUU-20 BOMB-ROCKET DISPENSER

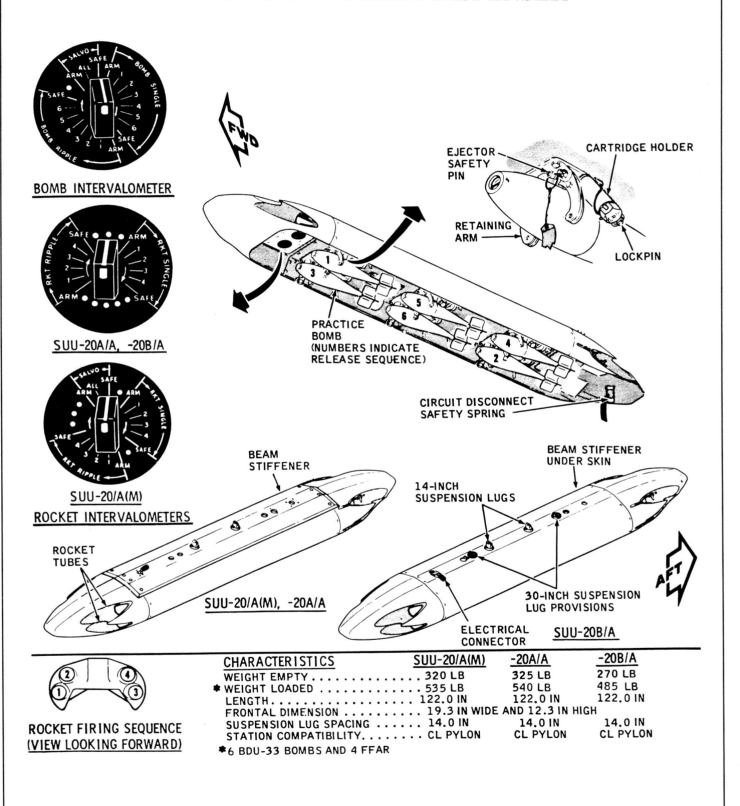

AF/B37K-1 BOMB CONTAINER

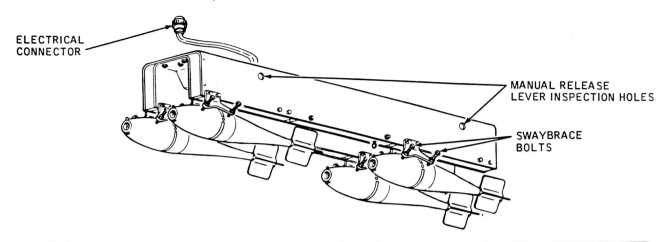

CHARACTERISTICS

```
WEIGHT EMPTY . . . . . . . . . . . . .  86 LB (APPROX)
WEIGHT FULL:
    WITH 4 BDU-33 BOMBS . . . . . . .  186 LB
LENGTH . . . . . . . . . . . . . . . . . . . .  46.0 IN
FRONTAL DIMENSION . . . . . . . . .  9.1 IN WIDE AND 6.0 IN HIGH
SUSPENSION LUG SPACING . . . . .  14.0 IN
STATION COMPATIBILITY . . . . . . .  CL PYLON
```

SUU-11A/A -11B/A GUN POD

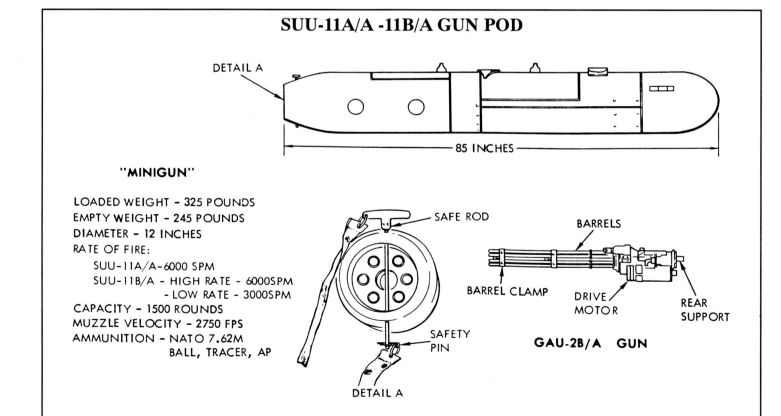

"MINIGUN"

LOADED WEIGHT - 325 POUNDS
EMPTY WEIGHT - 245 POUNDS
DIAMETER - 12 INCHES
RATE OF FIRE:
 SUU-11A/A - 6000 SPM
 SUU-11B/A - HIGH RATE - 6000SPM
           - LOW RATE - 3000SPM
CAPACITY - 1500 ROUNDS
MUZZLE VELOCITY - 2750 FPS
AMMUNITION - NATO 7.62M
           BALL, TRACER, AP

selector switch providing optional high (6000 SPM) or low (3000 SPM) gun firing rate. A single SUU-11 pod can be mounted on the centerline pylon. The gun pod is highly accurate, with minimum weight, and high reliability. The pod carries 1500 rounds of ammunition (15 seconds of firepower at the high rate of fire and 30 seconds at the low rate). The gun pod uses NATO standardized 7.62 millimeter ammunition. Expended ammunition cases are ejected from the bottom of the pod.

PERFORMANCE

MINIMUM CREW REQUIREMENT

The minimum crew requirement for the T-38 is one pilot. Solo flights must be made with the pilot flying the aircraft from the front cockpit.

THROTTLE SETTING THRUST

Normal Thrust

Normal (maximum continuous) thrust is the thrust obtained at 98.5% RPM or 630xC EGT, whichever occurs first.

Military Thrust

MIL (military) thrust is the thrust obtained at 100% RPM without afterburner operation.

Maximum Thrust

MAX (maximum) thrust is the thrust obtained at 100% RPM with the afterburner operating. Afterburner range extends from minimum afterburner of approximately five percent augmentation above MIL thrust to maximum afterburner, which is approximately 40 percent augmentation above MIL thrust.

AIRSPEED LIMITATIONS

WING FLAPS

The T-38 should not exceed the following airspeeds for the wing flap deflections:

1% to 45%	300 KIAS
46% to 60%	240 KIAS
Over 60%	220 KIAS

LANDING GEAR

240 knots indicated airspeed is the maximum limit for the landing gear extended and/or landing gear doors open.

NOSEWHEEL STEERING

65 knots indicated airspeed is the maximum limit for use of nosewheel steering.

CANOPY

50 knots indicated airspeed is the maximum limit for taxiing with a canopy open.

LOAD FACTOR LIMITATIONS

The following are the G limits for the T-38 in symmetrical flight:

Load Factor (G's)	Weight of Fuel Remaining (Pounds)
-2.4 to +6.0	Fully fueled
-2.6 to +6.4	2700
-3.0 to +7.33	1400 or less

T-38 TAIL CODES

AD	3246 Test Wing, Eglin AFB, AFDTC (T)
BB	9 WG/RW Beale AFB (CT)
CB	49 FTS (AT), 50 FTS (T), 14 FTW, Columbus AFB
CM	50 FTS, 14 FTW, Columbus AFB (T)
DY	96 WG, Dyess AFB/7 WG, Dyess AFB (CT)
ED	6510/412 Test Wing, Edwards AFB (T)
EL	28 BW, Ellsworth AFB (CT)
EN	88 FTS (AT), 90 FTS (T) 80 FTW, Sheppard AFB
ET	3246/46 Test Wing, Eglin AFB AFDTC (T)
GF	319 BW/BG, Grand Forks AFB (CT)
HM	49 FW, Holloman AFB (T, AT)
HM	479 TTW, Holloman AFB (T, AT)
HO	49 FW, Holloman AFB (CT, AT)
HO	479 TTW, Holloman AFB (T, AT)
HT	6585 TG, Holloman AFB (AT)
HT	46 TG, Holloman AFB (AT)
LA	2 BW, Barksdale AFB (CT)
LB	54 FTS, 64 FTW, Reese AFB (T, AT)
LL	3640 PTW, Laughlin AFB (T)
MO	366 WG, Mountain Home AFB (CT)
MT	5 BW, Minot AFB (CT)
OF	55 WG, Offutt AFB (CT)
OZ	384 BW/BG, McConnell AFB (CT)
RA	560 FTS, 12 FTW, Randolph AFB (T, AT)
SJ	4 WG, Seymour-Johnson AFB (CT)
TR	4450 Tactical Group, Tonopah Airfield (T)
TR	37 TFW/FW, Tonopah Airfield (CT)
VN	71 FTW, Vance AFB (T, AT)
WF	80 FTW, Sheppard AFB (T)
WL	82 FTW, Williams AFB (T)
WM	509 BW, Whiteman AFB (CT)
XL	47 FTW, Laughlin AFB (T, AT)

(AT) = AT-38B
(CT) = Companion Trainer
(T) = Standard Trainer

COLORS AND MARKINGS

Paint numbers are Federal Standards (FS) numbers

T-38A - White (17875)

LEAD-IN FIGHTERS
AT-38B - Blue 35450   Blue 35164   Blue 35109

COMPANION TRAINERS

UNITS	TAIL CODE	AIRCRAFT COLOR	MARKING COLOR
2 BW	LA	White (17875)	Black (37038)
	LA	Gray (36118)	Black (37038)
4 WG	SJ	Gray (36173)	Black (37038)
5 BW	MT	Gray (36118)	Black (37038)
7/96 WG	DY	White (17875)	Black (37038)
	DY	Gray (36118)	Black (37038)
9 WG	BB	White (17875)	Black (37038)
	BB	Black (37038)	Red (31136)
28 BW	EL	White (17875)	Black (37038)
	EL	Gray (36118)	Black (37038)
49 FW	HO	Gloss Black (17038)	Gray (36270)
55 WG	OF	White (17875)	Black (37038)
	OF	Gloss Gray (16173)	Black (37038)
319 BW/BG	–	White (17875)	Black (37038)
	GF	Gray (36118)	Black (37038)
366 WG	MO	White (17875)	Black (37038)
	MO	Gray (36118)	Black (37038)
384 BW/BG	OZ	White (17875)	Black (37038)
	OZ	Gray (36118)	Black (37038)
509 BW	WM	Gray (36099)	Gray (37200)

USAF AGGRESSORS
GHOST - Blue 35237   Blue 35622   Gray 36307   Gray 36251
GRAPE - Blue 35414   Blue 35109   Blue 35164   Blue 35622
SNAKE - Brown 30118   Yellow 33531   Green 34256
LIZARD - Brown 30118   Yellow 33531
GLOSS GRAY - Aircraft Gray 16473

4450 TEST GROUP/37 TFW

3 GRAYS - Light Gray 16440   Gray 16473   Dark Gray 16251